P9-AGW-192

BUILD YOUR OWN STONE HOUSE

Using the easy, Slipform Method

by
Karl and Sue Schwenke

GARDEN WAY PUBLISHING
CHARLOTTE, VERMONT
05445

Copyright 1975 by Garden Way Publishing
Charlotte, Vermont 05445

All rights reserved—no part of this book may be reproduced in any form without permission in writing from the publisher, except by a reviewer who wishes to quote brief passages in connection with a review written for inclusion in a magazine or newspaper.

Library of Congress Catalog Card Number: 75-16831
ISBN 0-88266-069-1 (pbk)
0-88266-071-3 (cloth)

Printed in the United States

Fourth printing, May 1977

Library of Congress Cataloging in Publication Data
Schwenke, Karl.
    Build your own stone house
    Bibliography: p. 151
    Includes index.
    1. Stone house—Amateurs' manuals. 2. House
construction—Amateurs' manuals. I. Schwenke, Sue,
1929-   joint author. II. Title.
TH4818.S75S38   690'.8   75-16831
ISBN 0-88266-071-3
ISBN 0-88266-069-1 pbk.

We dedicate this book to Helen and to Scott Nearing. Their example was the rock upon which this house was built.

Karl and Sue Schwenke

When I can hold a stone within my hand
And feel time make it sand and soil, and see
The roots of living things grow in this land,
Pushing between my fingers flower and tree,
Then I shall be as wise as death,
For death has done this and he will
Do this to me, and blow his breath
To fire my clay when I am still.

From *Collected Poems 1930-1960* by Richard Eberhart. Copyright © 1960 by
Richard Eberhart. Reprinted by permission of Oxford University Press, Inc.

Introduction

*M*essing about with stones is an addiction. For the stone addict, putting stones into a wall is an elemental act of life. To my way of thinking, there is absolutely no more constructive or more affirmative thing you can do than to build yourself a stone house. Consider . . . what single thing do you most want from life? Is it security? Could anything be more secure than spending the rest of your days sheltered by stone walls? Is it a monument to your existence? Every stone house is a monument to its builder. Is it the satisfaction of fulfilling a primal creative urge? What could be more creative than assembling order from chaos?

Like the Adélie Penguin's, my home is my nest — an agglomeration of stones gathered from hither and yon, (and, like the penguin, I have even been known to steal a stone or two from my neighbor). Penguins, however, are limited in their appreciation of stones. Man has always treated stone more imaginatively. It began when Glog-the-caveman picked up a stone and heaved it at his neighbor. That is the problem with a single stone. I mean, what else can you do with one stone but throw it? A pile of stones is a far more constructive thing.

There is a joy to be discovered in piling up stone for your house. You cannot find it in a book, not even in this one. The exhilaration of extending your body's muscles to their limit as you sweat under the loads of bigger and bigger rocks and endless shovelsful of concrete, is one that has to be experienced rather than described. The rekindled body seems to shrug off known threshholds of exhaustion. Heightened sensual awareness *accompanies every stage of physical exertion. Each stone that is added to the house, each level of slipforms that is removed — all contribute to a creeping sense of excitement and elation that swells and swells until each waking moment is suffused with an aching realization of completeness.*

But the joys of building a house of stone are not all physical. Overlying the whole process is the distinct intellectual awareness of control *— of having a firm hold on your own fate. This awareness permeates each working day with a satisfaction that is heightened by the basic nature of the earthy material with which you work. Building your own home is almost instinctual, and undertaking the task is as natural as bees and honey. In the final analysis, what* is *important is that you are doing it yourself — that you are living your own life, not paying others to do the living for you.*

Indeed, building a stone house is an addiction. When our house neared completion, I felt a sense of impending loss. I had been wallowing in my glory like a pig in a mud puddle, and now it was all going to end. The day inexorably arrived when there was not one more stone to lay, not one more form to set, and not a single nail left to be driven. What does the stone house addict do when confronted with Armageddon? He undertakes to build a stone barn.

Karl Schwenke
Vermont, 1975

"If you want to try a new way, you've got to build something."

The Last Whole Earth Catalog

Contents

Cross-Sections

Chapter 1

The fieldstone house is country — it does not pretend to something that it is not. Like diamonds in the rough its whole, natural stones connote honesty, utility and simplicity. Uncut, uncoursed and yet undeniably *right*, fieldstones combine in a farmhouse wall in a way that evokes a sense of completeness with the land.

These were the qualities that started us considering fieldstone for building. We had the ideal site; we liked the "feel" of stone textures; but we were inexperienced with building techniques.

The traditional methods of dry laying (laying stones atop one another without bonding agent so they remain in place by their own weight), and hand laying (the same as dry laying except that it uses mortar) required too long a time in construction. Therefore we finally settled upon the slipform method of building with fieldstone.

Simply stated, slipforming incorporates elements of hand laying, but the wall is laid up between forms and then is backed with a pour of concrete. With adequate reinforcing, this backing is both structurally superior to the other alternatives and is much faster in construction.

The process of slipforming — examined step-by-step in Chapter 5 — offered to us a minimal expenditure for forming materials, in that it incorporated using, reusing and reusing yet again the same standardized forms, all in the same wall.

The obvious economy of the method, the availability of free stone on our site, and the simplicity of the slipform technique — all contributed to convincing us that we should undertake the building of our first fieldstone house.

* * * * * *

It was a hot, cloudy New England day. Sweat rolled down my cheeks into squinting eyes. I was straining with a particularly stubborn stone, lowering it delicately into place between the forms, at the same time trying to ignore a black fly crawling on my eyebrow. To swat meant to drop the stone, and that might well destroy the whole tentatively-placed course I had just laid up.

"Ain't a *rightly* laid up stone house," boomed a strange voice behind me.

Startled, I shifted my grip a fraction. I felt the stone going and made a last, desperate clutch that only succeeded in trapping my fingers between the stones.

Forgetting my precarious toehold on the forms, I did a painful jig trying to free the imprisoned hand. My foot slipped, and suddenly there I was dangling helplessly by my armpits.

It seemed hours later when something solid was shoved under my thrashing feet. My rescuer proved to be the same stranger whose booming pronouncement had started the whole chain of events, and who now stood grinning foolishly as I rubbed my armpits and muttered "Thanks."

"Oh, t'warn't nothin'," replied the stranger. "Hangin' there like you was 'minded me of a hog with his foot caught in the stable floor come feedin' time."

Comment on my part was stifled by three factors that somehow got past my crushed fingers and chafed armpits: First, I couldn't think of a suitably crushing riposte. Second, my wife Sue arrived with words to soothe my wounded ego. And third, I realized the stranger was surrounded by a solid barrier of Southern Comfort. Introductions were made all around.

"Name's Charlie . . hired hand't the fawm up the road," he said. He was the first of the local people to visit our building site, and I resolved to be neighborly and civil.

"Nope," stated Charlie dogmatically, "Ain't a rightly laid up stone house."

I kept my voice even as I allowed how slipforming was a different way to build a stone house. I explained that slipforming was economical in its use of lumber, and that the process was a fast, simple and efficient method. Warming to my subject, I described how the forms were laid up, braced, wired, stoned and poured, and then I showed him the stones we were selecting from.

"You carry all them stone over hyah?" he asked as we picked our way between the piles. His eyes mirrored a patronizing tolerance usually reserved for wayward children and ignorant down-country folk. "I kin see y'gotta keep your courage up t'make one of these slapped dashed houses."

"Slipformed," I corrected him testily.

"Slip whatevered." He dismissed the subject as he wavered across a four-foot bridge over a one-foot trench. "Stone by stone — that's the way t'lay up a stone house."

Then squinting owlishly along the wall's length, alternating closing one eye and then the other, he pronounced: "She's leanin' out 'bout half inch at the top." Sighting down his extended arms he intoned solemnly: "Y'see, she's got a cant on 'er wus'n a lame heifer."

"You figure the wall is out about a half inch at the top, eh?" I asked politely, winking at Sue.

"Eeyup," he concluded with certainty. Then he made off, wandering down the driveway to his pickup.

We laughed about the visit after he left. But four weeks later, while cutting joists for the ceiling, we discovered that the joists were short — one half inch short.

* * * * * *

"It looks so . . . so unending," Sue said, looking over the proposed building site.

"Yeah. It seemed different on the plans we made."

Staring at the unbroken ground where we intended to build our stone house, the immensity of the project descended over us like a cloud. Flags on straightened clothes hangers marked the perimeter of the planned excavation. Their gay fluttering belied the gloom we were feeling.

"It seems impossible," Sue observed dismally. "It's so big. I don't even know where to start."

"At the beginning," replied Karl, rolling up his sleeves. "Then we break it down into smaller chunks that we can handle."

* * * * * *

The image of those two "young" people haunts us. Helen and Scott Nearing are never very far from our thoughts as we build our stone house. Grubbing about in the spring mud I recall their credo, "The good life is doing for oneself." Heretical as the thought was, I contemplated a contractor. My heresy was abetted by the swarm of black flies and mosquitoes that attacked us as we worked on our footing trenches.

Although we never met the Nearings personally, each shovelful of mud and water we throw out of the trench somehow brings them closer. Their sweat mixed with the mud too. Grinning to myself I reflected that Scott Nearing would have been counting the shovelsful as he threw them out of the foundation trench. Precise, that's what he would be. "One hundred ninety feet of perimeter trench, exterior measurement," he would say, "requiring eighty thousand, nine-hundred and forty two shovelsful." My shovel rang as it struck another stone. I swore and dug at a persistent black fly that had found its way into the top of my glove.

Sloshing about in the mire to where I could get another angle of attack, I wondered how many people even knew about the Nearings. How many had read their book *Living the Good Life?* How many knew that their lives had touched a wistful and responsive chord in the minds of thousands of equally "young" readers? My shovel struck the stone again, and I began to dig around it.

Now each dollop of mud stuck to the shovel's surface tenaciously as I tried to throw it aside. Digging a trench in stony soil is a humbling experience. I stopped for a moment's breather and watched Sue's progress farther down the trench. While she had not removed as much dirt as I had, it was progress. She was awkward, but persistent.

Companionable love is a curious thing. The Greeks were right in assigning many meanings and words to cover the various kinds of love. We had spent several hours working in this trench without exchanging a word. Words are often unnecessary, even dangerous to this kind of love.

"Honey," Sue interrupted my wanderings, "I've got another big bone of a rock." Her voice contained grains of discouragement and tiredness as she thumped her shovel against a large stone.

We had been digging continually now for a week, and the stony Berkshire loams of our farm had allowed us to penetrate only about eighteen inches. The first seven inches of topsoils had been easy, and then we had come to clay and large rounded stones that sometimes measured two or more feet in diameter. To put our foundation safely below the frost line would, at this rate, take us two more full weeks. This was two weeks longer than we had allowed for, and a prescient icy wind raised goose pimples on my sweating back as I projected this delay onto the other end of our short New England building season.

"The Nearings would never think of bringing in a backhoe to dig *their* foundation trench," said Sue in mock righteousness. She sometimes has an uncomfortable ability to candle the recesses of my skull. We grinned at each other ruefully as I climbed out of the trench.

"You had better clean the mud off your boots if you're going in to town," she said.

I looked down at the sticky clay that made my feet twice their normal size, and replied, "The clay befits a fallen paragon."

* * * * * *

Written credos are only valid as long as the ink is wet, but to our way of thinking everyone should write one now and then — if for no other reason than to use it as a mental emetic. Here then is/was our credo, somewhat abridged:

We are determined to involve ourselves as directly in the basic daily processes of living as we can. Shelter and food are two of the basics that we will try to provide for ourselves.

Our determination is born in the treadmills of the cities where we spend nine hours of each day working at jobs only distantly related to basics — basics that we pay others to build and grow for us. We have come to feel that the social and governmental fabric supporting these treadmills is so rotten and unresponsive to change as to warrant a re-examination of our participation. We find that we are not a constructive force in this fabric. In fact, we have decided that by our mere participation, we are contributing to a negative and destructive social force.

By way of solution, we now settle on our own private "revolution" — on beginning anew. We will attempt to keep as much self-righteousness out of this choice as we can, recognizing that there are potentially as many solutions to problems as there are those making choices. Our decision is taken with a fair share of humility — not to mention qualms. Two forty-year-old citified professionals attempting to be "reborn" sounds on first blush like a lot of utopian nonsense.

Our revolution entails finding a place where we can exercise a maximum of sovereignty over our day-to-day lives, and where we can shelter and feed ourselves. We seek a place where by our labors we can "do for ourselves" (with some obvious reservations) and, as a byproduct, better that part of the world we live in. We are no longer out to save the entire world, but will settle for improving that slice of it that we occupy. With us, we take as many tools as we can, and behind us we leave as much of the social facades and political sophistication as we are capable of shedding. Our aim is to live more simply and more creatively. A stone house is a natural outgrowth of this thinking.

* * * * * *

"Honey, I don't understand this," Sue said.

"What's that?"

"Well, we've got the footings and the foundations done, and now we're backfilling up to the grade line . . ."

"Yeah. So what's the problem?"

"We pour the basement floor next, right?"

"Yes, but we have to lay the drain pipe first."

"That's what I don't understand," she replied. "Where are we going to drain the drain?"

"Through the hole we left in the footing."

"What hole?"

"Arghh!"

* * * * * *

The most insensitive viewer cannot help feeling involved in a house made of stone. The natural walls exude a kind of warmth and a sense of comfortable well-being found in no other building medium that I know of.

Unlike the frame or other conventionally-built houses, the stone structure improves with age, its rough-textured walls welcoming the patina of colorful lichens and inviting the tentacles of climbing vines. It is also an ideal residence for bird life, protruding ledge stones, particularly up under the eaves, being especially inviting to summer visitors.

The clinching argument for stone is that it is a fine weather barrier — particularly to wind. It should be acknowledged that stone is *not* as good an insulator as wood (thickness for thickness), but this difference diminishes with increases in wind speed. Stone walls *should* be made thicker than wooden ones for structural reasons as well as thermal ones — a marriage of convenience.

Lastly, while stone transmits heat and/or cold more readily than wood, it has a thermal storage factor not available to many other mediums. Stone houses retain temperatures that they have been exposed to for prolonged periods of time. Thus the cooler night temperatures are extended over the major part of the heat of a summer's day, and the warmth of the late afternoon is carried over into the cool nip of the early evening. The net effect is to even out the extremes.

* * * * * *

A man-woman team must do more daily planning to build a stone house than, say, two men. Sue and I sought to share the day-to-day labor as equally as possible, but physical strength and the time pressure of an oncoming New England winter did limit the kind of work that Sue ended up doing.

Men are culturally imbued with a head start on all forms of physical work, and for Sue the first step was the rudimentary one of catching up on the basics of how to hammer a nail, saw a board, and shovel sand and gravel. We soon found that most men take this knowledge for granted, and it takes a careful and caring analysis to meet — much less resolve — the problem.

From the time he is a boy, a male is instilled with a basic awareness of body leverage that is not normally given (or offered to) the female. Starting with throwing a ball (or perhaps even earlier) a boy is directed toward functional body efficiency, and this acquired knowledge becomes a province of the automatic synapses — so much so, in fact, that he frequently errs in thinking that it is his masculine birthright.

During the building I brought home with me a young hitchhiker who professed to be an accomplished carpenter. Sue and I were in the midst of building forms, and I succumbed to the young man's good-natured insistence that he would be an asset. Fifteen minutes and fifteen bent nails later I realized that I had doubled my troubles. The young man reluctantly realized that carpentry skills were not his masculine birthright, and we parted a week later on good terms.

As a rule, women are not as physically strong as men, but they can do "a man's work." The answer lies in the time factor. Given enough time, any healthy woman (even though culturally deprived) can do the same tasks as her male counterpart. If this sounds patronizing to a stanch women's liberationist, it is not meant to be. For the two of us this problem, and the conclusions we reached, were eye-openers that we feel should be shared with other couples who may decide to build their own stone houses.

Our daily planning session (at breakfast) took these factors into account, and we set out what we hoped to accomplish that day. More often than not, we planned more than we actually managed to get done. Gathering materials, breakdowns of machinery, and fickle weather took their toll of our time, but these sessions were essential as they allowed us to make the most of our abilities, and at the same time to keep a sense of perspective on our over-all project.

"A time to cast away stones, and a time to gather stones together . . ."

Ecclesiastes 3: 1-8

"The real meaning of revolution is not a change in management, but a change in man. This change we must make in our own lifetime and not for our children's sake, for the revolution must be born of joy and not sacrifice."

Daniel Cohn-Bendit
Obsolete Communism, A Left Wing Alternative

Materials

Chapter 2

Jean Piaget once wrote, "We can know the true nature of an object only if we act upon it, break it down, and reconstruct it." So it is with a stone house. When you get right down to basic materials that go into it, you find: stone, sand and gravel, cement, steel and wood. The qualities of these basic materials, their availability, and their workability will affect the building process, and they will determine the structure's ultimate strength and its durability. Therefore, it is important to examine each of these materials carefully.

Stone has been used to shelter man since he lived in abandoned animal caves. Those caves that he hollowed out for himself remain as examples of man's first constructive efforts with stone — they were, in fact, stone houses by subtraction.

Being an ingenious animal, it did not take man long to find that he could reverse the process, and build with stones by addition. Using flat rocks, he put one atop another, and ended up with a wall — a structure which served to keep the chill winds away, and except for the addition of mortar, a structure which has not changed markedly to the present day.

The Stone Age ended 4675 years ago, but it would appear that this termination was premature, for as late as the beginning of the twentieth century stone, along with wood, was a predominant and basic building material.

Three fundamental facts were responsible for this longevity: 1) Stone is common — it is everywhere; 2) Stone is strong; 3) Stone is durable. These facts have not changed, but man's technology and values have. The rise of the technological age has produced an abundance of materials (like steel and plastic) that are easier to work with, stronger in consistency, and structurally more predictable.

Unlike stone, however, these technological materials require prodigious amounts of energy to produce, and it is only now (when we have been forced to balance energy input against actual value received) that these new technological materials have been called into question. Labor-intensive building mediums such as stone are once again viable — even desirable — alternatives.

Edward Flagg, a stone house builder of the 1920's and a pioneer of the formed stone house, pointed out a curious anomaly about the social acceptance of stone house: "Not so long ago, they were regarded as humble dwellings suitable only to pastoral circumstances. By contrast, today's house market, urban and rural, lists them at a premium. They are sought after by the status-conscious, and they command prices that are totally out of step with the market."

Social values and status symbols quite aside, we find that stone still makes a fitting farm house. To our way of looking at it, a stone house identifies with the soil in a way that no other structure could. We like to think of our stones as potential soil at rest midway between the mountain and the sea.

STONE

A stone's innate beauty should figure in any house building effort, but its functional nature as a supporting and enduring component in a wall are of primary importance. To determine a stone's functional characteristics, it helps to have a smattering of geological knowledge — but a discriminating eye and a measure of common sense play an even larger role. Building stones are generally classified according to their geological position, and physical structure. It is not within the scope of this book, nor the handiness of most prospective stone house builders to classify their stones by chemical composition.

As a general rule, the older a rock is, the stronger and more durable it is (this rule is laced with exceptions). Geological classifications divide rocks into groups: igneous (e.g. granite, greenstone, basalt), metamorphic (e.g. gneiss, marble, slate), and sedimentary (e.g. sandstone and limestone. While this kind of classification may help those with a working knowledge of geology, others may rely on a more pragmatic physical examination.

To get some idea whether your stone is a good building stone: First, heft the stone. A heavy stone is a dense one, and the odds are that it will be less likely to absorb water. This is important in cold climates where absorbed water that freezes can destroy the stone's surface and can eventually crumble it entirely. Feeling the texture of the stone will also help to determine the stone's absorption tendencies. The smoother the face, the more likely it is to shed water.

Secondly, bang the stone with a hammer. Get a first-hand experience in finding out how hard/easy it is to fracture it. Then, look at the inside of the stone carefully. Hardness (resistance to crushing) is one of the more desirable characteristics to have in your stone, and a close examination should give some clues to this particular stone's qualities.

Hard stones are characterized by the absence of fissures, empty cavities or loose laminations, and by the presence of a close granular or crystalline consistency. If the grains are coarse, but finely glued together, the stone is better suited to cold climes — it will resist exfoliation (cracking off of the outer layer).

Now, look at the fracture line where the stone broke. Is the fracture uneven — that is, does the fracture show sharp projections? If so, the stone is probably of granular structure. Crystalline structures show even fracture lines. A conchoidal fracture (with smooth concave and convex surfaces) is most desirable, since it is characteristic of a hard and compact structure.

The final test of a building stone is the most obvious of all. Hie yourself over to the nearest, already-built stone house to see whether it incorporates the same kind of stones that you anticipate using. The older the sample house, the better. Look closely at the stones in the wall to determine the kinds that have best withstood the tests of time, climate and structural needs.

Any discussion of the most common building stones should include gneiss, granite, marble, limestone, sandstone, slate and trap. There are, of course, several sub-categories of stone within each of these types, and fieldstone (a generic term for unquarried, natural stone) most often falls into one of these sub-categories. Unfortunately, the historically small-scale usage of fieldstone has never warranted formal studies, and there is a scarcity of statistical building information. Some salient facts regarding the more commonly used stones are summarized in table 2-1.

AGGREGATES

Aggregates come processed (washed, screened and graded) or as natural (bank run). Processed aggregates cost much more than bank run, but have the obvious advantages of uniformity, and, sometimes, cleanliness. You may, however, find that you have no viable alternatives, and will, of necessity, have to use one or the other. If you have to pay the price of processing, you should satisfy yourself as to the quality of the product. In order to make this assessment, you should know how it is graded, and what grades are best used in good concrete.

Processed sand is used in mortar and in concrete. Each grain of sand is a small stone that should be angular, clean and coarse. Most sand taken from lakeside or stream bank is rock meal that was ground in a

TABLE 2 — 1

Characteristics of
Commonly-used building stones

Rock	Average weight per cu. ft.	Durability	(natural)	Physical Charact.	Colors	Availability and other comments	Insulative Capacity
		Compressive Strength	Resistance to Water Absorption				
granite	167	Very Strong to Medium Strong	Resistant	granular, crystal-line	white, pink, grey, brown, black	Common, but mostly quarried. Good wall-stones	fair
gneiss	168	Very Strong	Resistant	granular, crystal-line, lg. layers, sometimes w/bands	white, pink, grey, brown, black	Most com-mon. Fine wall stones	fair
marble	170	Strong	Very Re-sistant	recrystal-ized lime-stone, mosaic texture, compact	red, pink, wht, grey brn, gold. Occ. mixed col.	Mostly quarried, nice threshold stones	poor
limestone	158	Medium Strong to Very Weak	Poor Re-sistance	loose crystal-line struct., sometime chalky	yellow, green, black, grey, white	Mostly quarried	poor
sandstone	139	Strong to Very Weak	Fairly Re-sistant	cemented sediments (mostly sand)	red, pink, white, grey, black	Regionally common. Wide var-iety. Fine wallstones	good
basalt	135	Very Strong	Very Re-sistant	fine grained, dark, igneous	black, grey	Regional, good build-ing stones	very good
slate	174	Strong to Medium Strong	Very Re-sistant	fine grained, compact sediments that layer and split easily	grey, black, green	Fairly common, fine roof or lintel stones	good

geological mill. It is composed chiefly of silica, with a dash of mica, hornblend, feldspar, etc. — it is the silica that makes for strength. After excavation, the sand is run over various grading screens. Satisfactory mortar sand (also suitable for concrete) can be graded over a screen having a ¼ inch mesh. Good concrete also needs the presence of some finer aggregates (10 percent of the sand should pass through a 50 mesh screen) so that the voids and interstices between the larger grains are filled. Lastly, well-processed sand is washed with water to remove the loam or organic material that would inhibit the adherence of cement to the sand particles. A simple, on-site test for the sand's cleanliness is to take a handful that is slightly damp. Press it. If it does not mold when the pressure is removed, you can figure that it is reasonably clean. If a mere nudge of the finger crumbles the whole mold, it will probably do for concrete, but would not do for quality mortar work such as that around your fireplaces.

Graded gravel undergoes the same processing as sand, but the meshes of the screening are larger. It is, of course, not used in mortar, and the size of gravel particles chosen depends upon what use the final concrete is to be subjected to. In walls, for example, where reinforcing rod is employed, the larger gravel should not exceed ¾ of an inch — this applies to the construction of most slipformed houses. Concrete for footings can, by contrast, use gravel up to 1½ inches in diameter, or even larger.

Bank run is usually considered too sandy for the making of good concrete, and the odds of it being contaminated cause it to be shunned by most commercial builders. Influenced by this judgment, we agonized at some length before finally coming to a decision. Our indecision arose from the fact that a neighbor's pit afforded a bank run that was gravellier than the usual, and it was considerably less in cost than the processed aggregated (the latter had to be hauled a considerable distance, and the firm's costs were unavoidably high).

Our final resolution to use our neighbor's bank run came after consultation with our bank balance, and considerable home testing. The latter, also described in some detail in the next chapter, entailed the handful-molding test mentioned above, and putting representative samples into suspension in a one quart glass jar. After shaking, the constituents of gravel, sand, silt (if any) settled out in layers, and we thereby determined some proportions. Sediments that settle on top of the aggregates should not exceed ⅛ inch depth for every two inches of sample. Our test, substantiated by later shoveling into the mixer, revealed the presence of spotty appearances of clay, but research indicated that clay contamination (up to 10 percent in some instances) did not materially decrease the strength of the concrete. Some specifications for mortar and concrete even allowed for as much as five per cent contamination of vegetative or suspended matter. We determined the absence of significant vegetable matter in our samples by adding one teaspoonful of household lye per ½ pint of water. Our test samples were clear, but had they been contaminated, it would have shown it by increasingly darkening shades of brown.

Chemical analysis can be made on bank run through a soils laboratory, and this would be the definitive way to assess bank run. The cost of this analysis was, in our case, prohibitive, and we relied upon our home tests and our common sense. The latter brought us to a compromise. For our footings and foundation walls we bought processed gravel to add to our bank run. Our rationale for using straight bank run for our walls was that we needed a more mortar-like consistency to our concrete so that it would tamp and flow between our wall stones. Whether by dint of this rationale, or by blind good fortune, our structures stand, and they show no particular weaknesses nor tendencies to crumble.

CEMENT

Poor concrete or mortar can also be affected by the proportion of cement to aggregate, and the particular characteristics of the type of cement used. Cement, as we refer to it here, is *portland* cement. No, portland is not a regional brand name that emanates from Portland, Oregon or Portland, Maine. It is so named because its color is similar to the color of the stone that is quarried on the Isle of Portland (just off the coast of England).

Historically, cementous material has been used for more than six thousand years, but portland cement did not appear until the early 1800's. It was then a mixture of limestone and clay, which underwent a processing that included pulverizing, proportioning, heating (to a clinker stage) and a final pulverizing. Today's cement uses some of these same ingredients (plus silica, alumina, iron oxide, gypsum, etc.), but the processing has become infinitely more complex. Today's initial pulverizing, for example, crushes the rock materials so fine that they will drop through a sieve which is capable of holding water. And, after the final grinding operation, each particle is so minute that 90 out of 100 of them will drop through a screen containing 40,000 openings per square inch. It is this fineness that allows the cement to coat each particle of sand and gravel, which in turn produces its excellent adhesive qualities.

When delivered, cement usually comes in 94 pound paper bags, and the bulk of the cement occupies one cubic foot of space. It is classed in five standard types (Types I, II, III, IV, V), each of which is designed for a specific purpose. Type III, for example, hardens and develops strength more rapidly — thereby allowing for quicker form removal. Type I, however, is the cement that most dealers carry in stock, and unless you specify otherwise, it is what will be delivered. We used Type I throughout all of our construction to date, but we are looking into the possibilities of using the relatively new process of air-entrained cements. This process involves either physically introducing air into the concrete, or using a cement containing chemicals, which, when wetted, increase the air content of the concrete. Reportedly, air-entraining will increase the concrete's workability, and its final lightness — with a minimum sacrifice in strength.

Unless you are building in an arid climate, storing cement presents problems. It is susceptible to moisture, and once wet (showing clumps that will not break when struck lightly with the back of a shovel), it is virtually useless. Finding a warm, dry place to put it during construction is sometimes impractical, and most builders make do by wrapping it in plastic — hoping that the formation of condensation on the underside of the plastic will not be enough to affect the top layer of bags. When possible, it is a good idea to share this storage problem with your cement dealer. You can do this by either buying the cement in smaller quantities, or by arranging with the dealer to buy in larger bulk units, with the provision that the dealer will warehouse the unused part of the order until you are ready to use it. Smaller lots at the construction site result in less damaged bags because they are much easier to protect.

STEEL

The use of structural steel in a stone house is a bonus to today's builder. As a reinforcement for concrete, it was virtually unknown until the mid-19th century, when an English plasterer named Wilkinson took out a patent. But it was still later, around the early 1900's, before the practice gained any real acceptance. Today structural steel and concrete are synonymous — and for good reason. Steel added to concrete can, in some situations, double or even triple the strength that would be gotten from concrete alone.

Stone houses, even those employing concrete between forms, were built without the use of steel well into the 1930's. But before you construe this fact to mean that steel reinforcement is unnecessary in stone house construction, you should first look at the number of them that show external signs of frost heaving or uneven settling — heaving or settling that has resulted in dangerous cracks or other structural disasters. Stone houses built in areas where earthquakes are a probability (or even a possibility) should *always* have walls reinforced with steel. This reinforcement should include: Vertically, ⅜" (No. 3) deformed (raised, bas relief patterns) rail reinforcing rod at 24" o.c., or ½" (No. 4) at 32" o.c.; Horizontally, ⅜" (No. 3) at 32" o.c. It is also wise in earthquake areas to provide expansion joints and other structural flexing mediums.

Most stone houses, however, do not require this kind of heavy reinforcement. By way of example, our climate, soils and structural needs indicated to us that we could get by with footings reinforced with three ½" (No. 4) bars at 7" o.c. (horizontally), and walls using ½" (No. 4) bars 8' o.c. (vertically). Our corners, where more stresses are at work, were reinforced more heavily (see Chapter 4).

When fretting over how much steel we should put into our footings, a trusted friend, who was schooled in architecture, pooh-poohed our propensity to "overbuild," saying, "The chances of an earthquake here in Vermont are about as probable as California sliding into the sea." One year later — almost to the day — Vermont experienced an earthquake, its first in nearly thirty years. Our friend should feel some tremors now (pun intended), as he has since left Vermont to live in California.

Geography, climate and unique structural needs dictate different reinforcement needs for each stone house. The odds are that most stone houses can be built with reinforcement plans similar to ours, but the reader can resolve any doubts by writing for the Building Code Requirements for Reinforced Concrete (ACI 318) of the American Concrete Institute, P.O. Box 4754, Redford Station, Detroit Michigan.

Some of the reinforcing steel that we used in our structures are not found in government reinforcement schedules. We have been known to use: sheep fence, tractor axles, barbed wire, railroad tracks and wire cabling that was used to cut gravestones. The use of these scrounged materials, however, means that you must exercise a little more common sense. Axles and railroad tracks, for example, do not show raised bas-relief patterns as does most reinforcing rod, and their tensile "holding" power is therefore considerably reduced. Most scrounged materials also have a tendency to be dirty and/or rusty, both of which are undesirable.

Recommended amounts of structural steel are often given a weight/foot ratio, and table 2-2 lists these ratios for easier calculation.

WOOD

Wood offers many alternatives for the house builder on a tight budget. New, kiln-dried lumber is, as a whole, easier to work with, but is far more expensive than locally-processed, air-dried lumber or that which one can salvage from another structure. Kiln-dried lumber is *not* necessarily better (or even drier) than locally produced, air-dried lumber.

TABLE 2 — 2

Weight/Foot Ratios for Standard
Reinforcing Rod

Bar Sizes		Weight in pounds per	Diameter in
Old (inches)	New (numerals)	foot	inches
¼	No. 2	.167	.250
⅜	No. 3	.376	.375
½	No. 4	.668	.500
⅝	No. 5	1.043	.625
¾	No. 6	1.502	.750
⅞	No. 7	2.044	.875
1	No. 8	2.670	1.000

We have had occasion to try all three alternatives. We found that salvaged lumber, while cheaper, required considerable effort to denail and size, and its splintery character caused us a lot of eye and marital strain. The eye strain resulted from unending sessions with a magnifying lens, straight pin, tweezers and band aids. Until you experience it, you can never adequately communicate the soul-satisfying sense of accomplishment that follows the final, bloody extraction of one of these minute slivers. The marital strain occurs when one of these "no-seeums" lodges stubbornly in your working hand, and after trying to get it out "wrong-handed", you turn to your mate for help. One must be desperate to resort to this. The utter callousness of the human race is indelibly pressed into your brain at that moment when, eyes alight with surgical fervor, your mate-turned-tormentor takes pin in hand.

Slivers aside, we found salvaged beams to be the best investment of time and money for the tangible results to be gotten. In building our house, we used hand-hewn beams taken from our old, collapsed barn, and used them as posts, girders and beams. Their well-tempered strength, and the sense of historical continuity that they lend, are qualities we can appreciate daily.

Structural members like joists and rafters make up most of the lumber needs for the stone house, and the most satisfactory source we have discovered for this lumber has been the trees on our own land. Like the fieldstone, this wood is native and seems to belong. We acknowledge that many (perhaps most) builders are not likely to have lumberable trees, but that should not dissuade them from exploring the alternatives. Stumpage (lumberable trees) can be bought from some private woodlots. If you are willing and able to do the consequent work of cutting the trees, brushing, and getting the logs to the mill — as we do — you can save a good deal of precious capital. Such an operation necessarily concludes with a storage time — a time when the lumber can dry. During storage the lumber should be carefully stacked with narrow spacers stuck between each layer in such a way that the air can circulate both horizontally and vertically. The rate of air drying varies depending upon the thickness of the pieces. The rule-of-thumb is to allow the stack to dry one year per inch of stock thickness, but this rule is only as good as the ventilation in the stack, and the vagaries of the season's temperatures.

After we have drawn our logs from our woodlot and transported them to the mill, we have the sawyer slash (plain) saw our framing lumber, and quarter-saw the boards that we use for sheathing of our floors and roofs. We have found that boards cut across the annual rings (quarter-sawn) tend to cup less than those that are slash cut. We have not found that cupping in two inch framing stock has proven to be a problem.

In general, wood is classified as either hardwood or softwood. Hardwoods are not considered here because they are used mostly in finish work, and thus outside the scope of this book. Softwoods are taken from coniferous (cone and needle bearing) trees, and they make up the bulk (81 percent) of the lumber that is commercially available. The major softwoods used for commercial lumber come from four geographical areas. Those areas and the trees for which they are noted are: 1) North: eastern white pine, red pine, jack pine, Canadian hemlock and spruce; 2) South: southern pines, bald cypress and coast white cedar; 3) West: western white pine, ponderosa pine, sugar pine, lodgepole pine, Engelmann spruce, larch, and Douglas fir; 4) West Coast: Douglas fir and California redwood.

Lumber that is made from these trees and processed commercially is graded on the basis of mechanical tests for stress, straightness of grain, freedom from imperfections such as knots, splits, stains, warps, pitch pockets, etc., and whether or not wastage can be expected when it is used. After testing, that which is intended for engineering construction (Structural Lumber), and that used for remanufacturing purposes (Factory and Shop Lumber) are set aside. Most house builders never even see these classifications of lumber. Yard Lumber is the appropriate classification for what is left over. It is the lumber you will find at the lumber yard, and it is broadly divided into three subcategories: "boards"

(1"-1½" thick), "dimension stock" (2"-5" thick), "timbers" (5" and thicker). Common boards are rated, No. 1 through No. 5. The latter is fit only for kindling and survey stakes. For floor or roof sheathing, the home builder will find No. 2 Common adequate. Dimension stock for joists, rafters and other framing confront the builder with further grading: Select Structural Construction, Standard, Utility and Economy. It is not wise to use the Utility or Economy grades of dimension stock for framing, as they are likely to have severe imperfections that could result in structural failure.

Like almost all building materials, wood has a high compression strength, but is weak when put in a tension situation. Each type of wood exhibits different strengths and weaknesses, and the builder confronted with a limited selection should acquaint himself with the particularities of the woods with which he must work. Structural tables given later in this book are based on an average tensile strength, and predicated on a good, clear piece of wood being used. Those desiring more span detail as it regards a specific kind of wood, should consult span data from testing services, or that compiled by the FHA (FHA Minimum Property Standards; No. 300; Nov. 1966; HUD, Washington D.C., 20410).

One final warning note should be included for the use of wood with concrete. When wood, without preservatives, is placed against concrete, it rots. This happens because concrete is hydroscopic, and in juxtaposition with wood, it provides an ideal growing medium for wood-destroying fungi. These fungi eat the cellulose of the wood — some even destroy the lignin. While some woods are more resistant to these fungi than others, they will all eventually succumb if conditions are favorable. The way to make these conditions unfavorable to decay is to treat the wood with a preservative that is either oil or saline in base. We used the oil based preservatives, but we have been given to believe that the water soluble salts are equally effective. Most of these preservatives are toxic, and should be treated with care.

> *"If you can pick it up, it's a nice stone — If you can't,*
> *it's a goddam boulder."*
>
> *Anon farmer*

> *"I wish I were a little rock, a sitting on a hill,*
> *A doing nothing, all day long, but just a sitting still;*
> *I wouldn't eat, I wouldn't sleep, I wouldn't even wash —*
> *I'd sit and sit a thousand years and rest myself, b'gosh!"*
>
> Frederick Palmer Latimer
> The Weary Wisher

Siting

Chapter 3

C hoosing to build of stone is a decision that is all too frequently made on esthetic grounds alone. If this is your situation, we suggest that you re-examine your priorities. There are pitfalls and problems to siting a stone house that you must confront if you want the building process to be a pleasant one, and the house to stand. Stone, for example, is a heavy building material, and if it is not *readily* available to the building site, its transport can become a nightmare. In assessing our home site, we confronted an old, randomly piled stone wall that separated the house site from our prospective garden site. We determined that we would have to remove it, and it seemed good horse sense to use that stone in our house wall. On its face, it appeared an economical use of effort as it would require a single move. Little did we know in those early planning stages that we would end up moving each and every stone at least twice, and more often three or four times. The cumulative poundage of stone we hefted on that first house amounted to nearly a thousand short tons, and that back-breaking figure does not include the weight of materials for footings, foundations, nor concrete for backing the walls. This weighty figure will daunt the dilettante, and well it should. Stone house building is not for the dabbler.

Now that the scary part is behind us, those who are still reading this will be comforted to know that about one half of that old fence wall that we aimed to use up in building our house still remains. Stone houses require far less actual stone than you would suspect. In adding up the stone we actually put in the walls, we came to the amazing conclusion that it would all fit in ten, big dump truck loads. The trouble is, most prospective stone house builders do not have this kind of heavy equipment, and the cost of renting or buying it is prohibitive. The odds are that if you are reading this book you too cannot afford to buy your stone, and if you could afford to buy it, you couldn't afford to have it delivered.

Many stone house builders we have come to know who lacked

available stone, have come up with unusual solutions. One young couple that commutes to a daily job told us that they would return each night to their house site with their small car packed full of stones they had picked up along the roads. They told us that they studiously avoided razing old stone walls, preferring to honor the owners' rights and the walls' historic integrity. Over the years, their growing cache of hand-picked stones has grown to the point that at this writing, they are well underway with their house. Another, of the "scrounging generation," haunts the dump of a nearby quarry, and with the quarry owner's permission (and encouragement) carts away stone by the pickup load. Yet another took a page from the book of the farmers who originally built the stone walls on his land. One wall is located in the encroaching woods, at some distance from his building site. With oxen and a stone boat (a sturdy metal skid), he is moving that wall, stone by cherished stone.

Contrast this builder's love for stone with the modern, "agri-businessman" farmer's attitudes. Today's farmer looks upon stone as a nuisance — a field impediment that entails a lot of sweat, blasphemy and good humor to banish to a stone fence wall. The farmer watching these stones moved again — even for the construction of a house wall — cannot believe his eyes. At best he views the procedure with amused tolerance, but he occasionally expresses alarm for his neighbor's sanity.

It is a mind-boggling rarity to see a stone house in this vicinity despite the ubiquity of the medium. This rarity flies in the face of the New England Yankee's proverbial appreciation of durability, economy and strength. One could understand the contradiction if, say, the land were of a swampy or clayey nature, but it is not. Good sandy loams and ledge underlayments, such as characterize our area, make it an ideal place to site stone structures.

Unfortunately many would-be stone house builders fail to assess their site properly, both for potential problems and their site's advantages. Their failure to assess the potential problems usually starts with the soils they intend to build on. Since the soil is the true foundation of any house, it is imperative that you get to know it better. Some soils will carry a heavier load than others (see table 3-1). Clay, for example, is not a friendly soil for stone houses, and building on wet, soft clay is courting disaster.

The first step to finding out what kind of soils make up your building site is to contact the nearest office of the Soil Conservation Service. Where possible, they will provide you with a soils map, and more often than not they will extend other personalized services and information at no cost to you. Over forty five percent of all private lands in the United States have been surveyed, and with a minimum of effort,

TABLE 3 — 1

Type of soils	Recommended maximum load per square foot
Soft clay	1 ton
Wet sand or firm clay	2 tons
Fine, dry sand	3 tons
Hard, dry clay or coarse sand	4 tons
Gravel	6 tons
Hardpan, or shale	10 tons
Solid rock	unlimited for domestic dwellings

one can get a pre-packaged appraisal of his site's soil type(s) and its general characteristics. The costs for these maps are negligible, and in many instances they are free. Some of the things you may learn from these maps are: ground loading potential; soil permeability (this gives data about potentials for basement flooding and information regarding the type of septic system your soils are best suited for); flood plain information (is your site a waterway for heavy storm runoff?); and soil growing capabilities (these are facts you will want to know after the building is done, and you want to cover the scar of the excavation).

While important, researching on maps and through books is not enough. Paperwork should be followed by a hard-working day on the site with a shovel and a pick. You should first look the site over, carefully assessing the slope, unusual variations in vegetation (possible signs of a stone or ledge close to the surface), and for obvious problems such as erosion marks, large boulders or potential, iceberg-like ledge formations. This exploratory day should be followed by some serious digging in the more dubious areas. If after going through these steps, you are still unsure about the soils type and its composition — particularly as to the percentage of clay and sand, you may want to try the "bottle test."

To make the bottle test, find a spot on the building site that appears to be somewhat representative of the whole. Dig down deep enough to avoid the sod or humus layer, and extract about a pound of the dirt. Remove all pebbles larger than your thumbnail, and then pick out any remaining vegetable matter. Put the remainder into a quart bottle, and then fill the bottle with water and shake for a few minutes. If the bottle doesn't break, you removed all the large pebbles as suggested above, and you will be rewarded by being able to watch the separation process that will follow. As the turbidity decreases, you will find the remaining pebbles and the coarse sand on the bottom of the bottle. On top of this layer will be finer sand, and on top of all will be the silt, clay and organic matter. Any turbidity that remains in the water will be very fine clay, which will settle if you leave it long enough. The proportions of this layering will give you an idea of the composition of your soils, and although the test is rough, it should suffice for most building sites.

But if you are still in doubt, you may want to send a sample away for a laboratory test. To do this you should contact your nearest State Agricultural Experimental Station, or your local County Agricultural Agent. For a minimal fee ($3—$10) they will provide you with sterile soils containers for your samples, and will even occasionally provide you with an auger for taking them. If you do not have an auger be sure to follow the alternative directions for taking samples very carefully (a drop of sweat from your hand, or an ash from your pipe can, for example, invalidate the whole process). In filling out the forms that accompany the samples, stipulate that you would like to build a house on these soils. Besides testing your house site, you may also want to include samples of the soils from the possible site of your septic system (clearly indicate to the laboratory which samples are which).

Part of siting your house necessitates determining where (if?) you can install a waste disposal system. A septic tank may be located as close as ten or twenty feet from the house, but a seepage system should be farther away. Both must be located downhill from the house site. To determine your soil's ability to absorb seeping effluent, you should continue your exploratory digging and make three or four percolation test holes. These holes should be four to twelve inches wide, and two to three feet deep. They should then be bottomed with two inches of fine gravel or coarse sand. After you have dug the hole, carefully fill it with clear water, and then keep it full for at least four hours. This saturates and swells the soils so that the timed test reflects more accurately the natural circumstances during the wettest times of the year. Next, adjust the depth of the water to about six inches over the gravel, and then time the diminishing water level for thirty minutes. The percolation rate is equal to the distance the water drops (in inches) divided into the time (in minutes). This rating may be used to determine: 1) whether the site is suitable for a drainage or seepage system, and 2) its required size. It is wise to make three or four separate measurements in each hole to reach an average. A rating figure of from 1 to 10 indicates an adequately porous soil for most systems, and you can go on to other siting problems.

While you are still scratching around the soils of your property, you may want to do a little treasure hunting for your own supply of sand and gravel. Yes, that sand and gravel you buy at the building materials store comes from the ground. You may strike it lucky and find these materials on your own land. Almost all "dirt" has sand and gravel in it, but it can also have undesirable things in it like clay or vegetable matter. To check for the latter you should use the bottle test described above, and add one teaspoonful of household lye to each one-half pint of water. If the water turns dark, you have vegetable matter in your sample, and it is likely that the gravel would have to be washed before use in construction. The final assessment, however, entails a more detailed analysis — an analysis that is made in Chapter 2. One last word of caution: local ordinances frequently codify or limit (and sometimes prohibit) the taking of sand and gravel. Perhaps you should dig into the ordinances before you dig into the land.

Using mortar or concrete in construction requires water at the building site. This could be construed to be an argument for building of stone, in that one must of necessity first find his water. A nearby neighbor constructed a beautiful, distinctive and innovative house on a hilltop, where, he belatedly discovered, there was no water. Extensive (and expensive) drilling efforts gave him no water, and he is now contemplating other equally expensive alternatives. Ours was a fortunate case, as we moved to an abandoned farm where a spring well was already established. Through inquiry we learned that our spring was "proven," in that it had continued to provide water through several severe drought years. In the country, we learned, you cannot readily plug in to a ready-made city water system. But we also learned that rural circumstances usually afford easier access to unpolluted waters of the surface variety. More importantly, once the system is going, there are no municipal waterbills to pay. Water for concrete or mortar should be of drinking quality, and should be dependable enough to provide you with 150 gallons per day.

How do you know whether you have an adequate supply of drinking water? Barring the possibility that you are an experienced well driller or a certified dowser (assuming that you subscribe to the verity of this ancient craft), you will probably have to seek outside help. Depending on your locale, water may be abundant a mere foot below the surface, you may have a surface run-off, or it may be caught deep in a subterranean aquifer. If there is no apparent surface water to be had, you will have to drill for it, and whether you do the drilling yourself or have it done, you should await the results before settling on a site. In either case you will have to await a time when you can accurately measure the quantity, and have the quality of it tested. Professional well drillers will give you an accurate rate of flow *once the well is in*, but the reputable driller will not guarantee you a specific flow before breaking ground. Because well drilling is (technology notwithstanding) a gambling game, the best that a driller can do is to give you some odds — based on scientific guesswork. You should be forewarned that most well drillers "copper their bets" by working on your capital — that is, you pay them for their work, whether they bring in a productive well or not!

In the region where we live, the typical builder "witches" for his water with a freshly cut forked twig — usually of a swampy variety of wood. He then "tests" his witching rod with a pick and shovel, or with a backhoe. Nearly all of our neighbors boast water sources obtained by this ritualistic method, and a jam-packed, yearly gathering of the National Water Dowsers Convention only thirty miles from our farm attests to the local respectability of the dowsing rod. There are, however, more skeptics than there are believers, and at the risk of changing a serious workday into a springtime lark, we would suggest that you try dowsing for yourself. A choice of a divining instrument is the only complication, and we will take that burden off your shoulders by recommending that you start with an eighteen-inch length of 1/16-inch brass welding rod. Bend four inches of the rod into a ninety degree angle making an "L" shape. Next, drop the short, four-inch length into a five-inch length of pipe (half-inch plastic pipe will do), and *voila* you have a dowsing rod.

To use the rod, walk slowly forward with the rod held before you in such a manner that the long leg of the "L" extends forward. You will be holding the pipe in your hand, which will in turn encase the brass rod. The idea is that your hand will not be touching the rod. When your walking progress takes you over water, the dowser claims that the rod will (if you have the ability) swing to the side.

Once found, water must be tested for quality. To do this, samples for laboratory analysis must be taken. Generally this is done throughout the States by the State Health Department. It is best to take this sample before and then after the well or spring has been developed — after it has had a chance to "settle down." Before taking your sample you should ascertain that waste systems are located at a sufficient distance that they do not affect your water. Local codes should be consulted to determine an acceptable distance, but a safe rule of thumb to use while siting is to keep all potential contamination sources a minimum of 150 feet away from your drinking water.

Should you plan to mix your own concrete, you may want to entertain the idea of doing it by hand or with a hand-powered mixer. But if you plan on doing the mixing with the aid of electricity, you should satisfy yourself that the costs of getting the electricity to the site are within your budget. Rural circumstances entail distances, and distance is a cost factor to your local power company. If there is any one thing that you may count on in this world, it is that these costs will be passed on to you. Assuming that you do not have your own power source, the problems that may arise from acquiring electricity may well alter your siting plans.

Other potential site problems turn around the weather. Wind, sun and precipitation each present unique siting problems that require the builder to familiarize himself with his local circumstances. With a little advance planning, however, one can turn a siting weather catastrophe into a building plus. For example, exposed, windy sites can be made into a "windfall" by taking advantage of one of the various wind-powered generators. You might even consider installing an old-fashioned windmill to provide your water.

Sunshine is usually never considered a problem, but if you find that you have an overabundance of this medium, you may consider the possibilities of putting it to work for you. Solar heat and power is a rapidly-growing aspect of house construction, and numerous plans and suggestions can be found in the bibliography.

Precipitation is always a problem in siting a house. There is nothing quite so miserable as a leaky basement, particularly if the problem could have been avoided with some advance planning. A fundamental problem most builders seem to share is to acknowledge that water runs downhill. Siting a house at the head of a "draw" or flood gulley seems a foolish thing to do, but it is done daily. Familiarizing oneself with local conditions can be done by the simple expedient of talking with neighbors who appear reasonably observant. Or, if you like to approach the problem more scientifically, you can check with the nearest office of the U.S. Weather Bureau. Seasonal flood conditions should be assessed for both maximum rainfall potentials, and for typical snow buildups. When melted, snow contributes enormous quantities of water (an inch of snow, for example, falling evenly on one acre of ground is the equivalent of about 2,700 gallons of water).

Stone houses can stand — even thrive on — most kinds of weathering, but frost or earthquakes can undo them. The reasons for this are fundamental engineering ones: Stone walls are "compressively" strong, but "tensilely" weak (unless reinforced with reinforcing rod or other steel). These engineering principles are more extensively discussed in Chapter 4, but simply stated compressive strength is the ability of stone or concrete to support direct loads without crushing or breaking. Tensile weakness is stone or concrete's proneness to break on the tension side (the side opposite the load) when under load, and when unsupported or reinforced. The earthquake problem is predominantly a localized one, and should you choose to build in such an area, you should be prepared to pay a larger bill for reinforcing rod. Further, you should familiarize yourself with expansion joints and other architectural "flexing" design methods. In siting your stone house in areas where there is a deep frost, you should plan on putting your footing below that frost line (as deep as four feet in northern parts of the U.S.).

Forseeing pitfalls and problems is an inherent part of stone house building, and solving these problems is for many people the most satisfying part of the process. But I would personally prefer to consider the pleasures of siting a stone house. Esthetics play a large part in the pleasures of siting. Daydreams are important here, and while walking around your building site is an ideal time to give them free rein. Picture in your mind's eye how you see yourself living on this site. Where, for example, do you see yourself working, playing, loafing, reading, being alone?

Like many people, we started our siting plans with some rudimentary ideas of attaining solitude, and filling every window with panoramic views. Time and a little more research altered some of our priorities, and added others. Solitude, for example, means something different to everyone who seeks it. Our rural circumstance is such that nearness to other people is not a present-day factor. However, our original plan was to site our house quite a distance from the road, and it was only after spending a winter clearing a long driveway of snow (the winter that preceeded our building), that we compromised. We moved the house site nearer the road. We settled on 300 feet, which was still a long driveway to be cleared, but the decision was in line with our new-found priorities. We are conscious of traffic on our road, but not unduly so. People in our neck of the woods are infinitely curious, particularly about a stone house, and we consequently have our share of summer "drive-bys" and "drop-ins."

How other people see your house — or, to put it another way, what your house contributes to the local community — is important. To self-righteously assert "that you don't give a damn what other people think" is nonsense. Like it or not, you are a part of your community, and you cannot resign from it by merely building a house. Communal participation aside, however, the way you view the house is, rightly, your primary concern. This was our reasoning as we considered where to site our house, and we settled on placing it in such a manner that the views of it we found *most* pleasing were available to *us* as went about our daily chores. These "outside-in" views involved, for us, siting and architectural considerations that revolved around our daydreams. We dreamed that our house would rise unobtrusively from the landscape, and by its nature and placement connote a sense of organic wholeness with its setting. We pictured ourselves returning at sunset from our fields to a house that welcomed us with a sense of unshakable domestic solidity. To make these dreams a reality, we nestled the house into the hillside contours, faced it so that the longer, lower lines were daily available for us to enjoy, and lowered the roof line as much as design would permit so that the house would blend into the rolling farmscape.

As we spend a good deal of our lives outdoors working about the farm, these views were of vital significance to us. But of equal import to us was how we "framed" the world to appear to us from the inside looking out. In siting we sought to take advantage of both the natural views and the views that we created (and are still creating). By far the cheapest to acquire when siting, are the natural views. These need not be of the sweeping, mountaintop variety — certainly ours is not. A view may well be confined to a single tree, garden, or (as in our case) a long, hardwood-fringed valley. Man-made views are generally longer in coming, and require careful planning. Landscaping, courtyards and compounds require several years to bring to anything resembling your dream, and you must be prepared to live with the transition period.

Views from the inside looking out are very personal ones, and preferences will vary. Your dream may involve taking advantage of a panorama, a body of water, a topographical feature such as a mountain, or the "view" may even be designed to bring in the sounds of a nearby brook. A neighbor friend whose house overlooks a sweeping, fifty-mile stretch of the Connecticut River laughingly points out that his house's most popular views, by family concensus, are the ones from their dining room window — of their active birdfeeder. As preferences are liable to change, so are the views that are dependent upon other people's ownership and whims. It is because of this that one should assign some priorities, and at the top of the list should be the natural phenomena, such as sunrises, sunsets and moonrises. Besides being reliable, they are also free.

The ultimate reliable bargain is sunshine. In climes where heat is a necessary factor, consideration of house placement is of prime importance. Where too much sun is a problem, orientation of the structure, by placement of windows and adjustment of the length of the shading eaves, will assure you maximum coolness during the hottest part of the day. In areas of extreme seasonal variation, the house siting should take advantage of sunlight for heat during the winter months, and allow a minimum amount of sunlight to heat the house during the midsummer months.

Our house is located in a region of extreme temperature fluctuations, and we endeavored to meet this problem by constructing a small, cardboard scale model of our house. We took particular pains in building the scale model to see that the proposed eaves were in scale. Then by using a solar table (see table 3-2) and a carefully placed floodlight, we calculated how long our eaves needed to be in order to provide maximum shade in midsummer, and maximum sunlight inside

TABLE 3 — 2

Table showing the azimuth* and altitude** of the sun

WINTER (Dec. 22)				FALL (Sept. 23) (Mar. 21) SPRING				SUMMER (June 22)			
AM	PM	Az.*	Al.**	AM	PM	Az.*	Al.**	AM	PM	Az.*	Al.**
				25° North Latitude							
No	on	180°-0'	41°-30'	No	on	180°-0'	65°-0'	No	on	180°-0'	88°-30'
10:00	2:00	146°-30'	33°-30'	10:00	2:00	126°-0'	51°-30'	11:40	12:20	107°-0'	85°-0'
8:00	4:00	125°-0'	14°-30'	8:00	4:00	103°-30'	27°-0'	11:00	1:00	93°-0'	76°-0'
6:50	5:10	116°-30'	0°-0'	6:00	6:00	90°-0'	0°-0'	8:00	4:00	78°-0'	35°-30'
								5:10	6:50	63°-30'	0°-0'
				30° North Latitude							
No	on	180°-0'	36°-30'	No	on	180°-0'	60°-0'	No	on	180°-0'	83°-30'
10:00	2:00	148°-30'	29°-0'	10:00	2:00	131°-0'	48°-30'	11:40	12:20	144°-30'	82°-0'
8:00	4:00	126°-0'	11°-30'	8:00	4:00	106°-0'	25°-30'	11:00	1:00	112°-30'	75°-0'
7:00	5:00	117°-30'	0°-0'	6:00	6:00	90°-0'	0°-0'	8:00	4:00	81°-30'	36°-30'
								5:00	7:00	62°-30'	0°-0'
				35° North Latitude							
No	on	180°-0'	31°-30'	No	on	180°-0'	55°-0'	No	on	180°-0'	78°-30'
10:00	2:00	149°-30'	25°-0'	10:00	2:00	135°-0'	45°-0'	11:00	1:00	127°-3(	72°-30'
8:00	4:00	126°-30'	8°-30'	8:00	4:00	108°-30'	24°-0'	10:00	2:00	105°-30'	61°-30'
7:10	4:30	119°-30'	0°-0'	6:00	6:00	90°-0'	0°-0'	8:00	4:00	85°-30'	37°-0'
								4:50	7:10	61°-0'	0°-0'
				40° North Latitude							
No	on	180°-0'	26°-30'	No	on	180°-0'	50°-0'	No	on	180°-0'	73°-30'
10:00	2:00	150°-30'	20°-30'	10:00	2:00	138°-0'	41°-30'	11:00	1:00	138°-0'	69°-0'
8:00	4:00	127°-0'	5°-30'	8:00	4:00	110°-30'	22°-30'	10:00	2:00	114°-0'	60°-0'
7:30	4:30	121°-0°	0°-0'	6:00	6:00	90°-0'	0°-0'	8:00	4:00	89°-0'	37°-30'
								4:30	7:30	59°-0'	0°-0'
				45° North Latitude							
No	on	180°-0'	21°-30'	No	on	180°-0'	45°-0'	No	on	180°-0'	68°-30'
10:00	2:00	151°-30'	16°-0'	10:00	2:00	141°-0'	38°-0'	11:00	1:00	145°-30'	65°-30'
9:00	3:00	139°-0'	10°-0'	8:00	4:00	112°-0'	20°-30'	10:00	2:00	121°-30'	57°-30'
8:00	4:00	127°-30'	2°-30'	6:00	6:00	90°-0'	0°-0'	8:00	4:00	93°-0'	37°-30'
7:40	4:20	124°-30'	0°-0'					4:20	7:40	55°-30'	0°-0'
				50° North Latitude							
No	on	180°-0'	16°-30'	No	on	180°-0'	40°-0'	No	on	180°-0'	63°-30'
10:00	2:00	152°-0'	12°-0'	10:00	2:00	1143°-0'	34°-0'	11:00	1:00	150°-30'	61°-0'
9:00	3:00	139°-30'	6°-30'	8:00	4:00	114°-0'	18°-30'	10:00	2:00	127°-30'	54°-30'
8:00	4:00	128°-30'	0°-0'	6:00	6:00	90°-0'	0°-0'	8:00	4:00	97°-0'	37°-0'
								4:00	8:00	51°-30'	0°-0'

*Azimuth: This is the angle measured horizontally from the North meridian (a compass reading).

**Altitude: This is the vertical angle, measured between the sun and the horizontal plane of the horizon. It should be noted that the North meridian is the *true north*, not magnetic north.

the house in midwinter. Our success may be determined by the fact that in midwinter our entire floor is bathed in its warming glow (twenty four feet, wall to wall), and the midsummer sun never encroaches beyond the edge of the windowsill.

But the pleasures of solving these kinds of siting problems is merely the appetizer to building a stone house. Getting to the building process is the main course, but you must first make plans and estimate your costs.

"Build of your imaginings a bower in the wilderness ere you build a house within the city walls."

Kahlil Gibran
The Prophet

* * * * * *

Dowsing: "People also can cultivate a sensitivity to very slight magnetic field changes. This is the trick of the dowser, who senses underground water pools through exceedingly small changes of magnetic field strength. In 1962 a Sorbonne physics professor, Yves Rocard, discovered that by holding the arm very taut, and balancing a long stick, the nerves and muscles in the arm would become sensitive to small changes of magnetic field strength. These changes, from water in the soil would cause the muscle to relax and the stick to dip . . ."

Gay Gaer Luce
Body Time

Planning & Estimating

Chapter 4

B y now you have a fair idea of where you want to site your home, and all that remains is to design the structure itself. So what do you do now? Do you take one look at all the engineering details, recoil, and then throw it all in some architect's lap?

The deification of expertise has, over the years, cowed the average home builder to such an extent that he has virtually given over his innate abilities to problem-solve. To our way of thinking it makes no sense at all to have an architect satisfy his personal creative fantasies at your expense. A builder designing his own home is acting out his *own*, direct personal experience — an architect specialist would, at best, regard the design of your home as an intellectual exercise. For us an architect was, and is, a valuable resource person. We do not hesitate to consult him with specific engineering problems or other detail, but we would no more entrust the designing of our home to him than we would allow him to prescribe how we are to live out our remaining years. Designing your home is the opportunity of a lifetime, a chance to make *your* fantasies a reality.

You *can* do your own planning and designing for your stone house, but before you begin you should check your local building codes. In general you will find that the further you are from urban centers, the less complicated the permit procedures. Farming areas like the one we live in are, for example, concerned only with waste disposal and proximity of buildings to property lines.

A slipformed house occupies a unique place in building codes. It does not fit into the categories of a concrete walled house, nor of a random, fieldstone (rubble) construction technique. Most codes, for example, indicate that a builder may raise an unreinforced concrete wall with a width that is 1/20 its overall height (providing soils conditions and other structural limitations are met). These same codes stipulate that a traditional, hand-laid stone house, employing random patterned fieldstone (rubble), should have a wall that is a minimum of 16 inches thick. We compromised. Our walls average 12 inches in thickness (foundation wall is 14 inches, upper exterior wall is 10 inches, while our average height of wall is only 13 feet. Finally, our compromise extended to using greater amounts of steel reinforcement than codes called for in an all-concrete wall of lesser width.

The first step you should take in planning your house is to draw some floorplans. You do not have to be an artist to do this, as these sketches *should* be rough and not bound by the inhibitions of straight edges or other drawing paraphernalia. Floorplans are one-dimensional plan-view drawings whose purpose is to give you a feeling of the spatial relationships within the house, a sense of the house's traffic flow, and a chance for you to visually work out your own use patterns. They should be undertaken with a sense of freedom, a private agreement with yourself that at the outset anything goes. The brass tacks of cost estimating, engineering, weather and other mundanities will make their inroads soon enough.

For us, floorplans occupied many a winter's evening, and they consumed copious quantities of scratch paper. We began by imagining ourselves living in our house through the course of an ideal day. This day saw us cooking our meals, eating, working (both indoors and out), loafing, bathing, reading, being alone, entertaining company and sleeping. By this process we began to allot priorities according to our needs and wants, and we came to the interesting (to us) conclusion that what we really wanted was *two* houses. One house we called the "social house," and the other we named, "the retreat." Our social house incorporated cooking, eating, loafing, reading and entertaining company, and our retreat was for working, bathing, reading, being alone and sleeping. We hooked the two "houses" together by a corridor that was flanked by work and loafing rooms, and thereby assured the privacy and separateness of our retreat.

Floorplanning can be a fascinating experience for those who have never attempted it. If you have let yourself go as was suggested above, you will have explored all manner and shapes of houses. You may have drawn conventional rectangles, "T" shaped houses, "E"'s or "H"'s. You may have gone into the faceted circle houses, making pentagons, hexagons or octagons — some of the latter may have surrounded a courtyard.

Even as you are making some of these preliminary drawings — few of us end up with what we started with — you begin to see problems emerging. You may, for example, have found that you had to put in a traffic corridor that used up valuable interior space, or you may have found that there were not enough south-facing walls on which to situate activities where you wanted the blessings of sunlight. It is in these problems and the resultant solutions that the value of this stage of the planning makes itself known. Remember that it is always easier to move a stone wall with a pencil than it is with a jackhammer.

Generally in your floorplans you should attempt to make your house's floorspace as large as possible. If you stretch your building budget anywhere, this is the area where it should be done. Other general suggestions for floorplanning are: 1) Keep the outside facets of the house as simple (straight) as possible; 2) Strive to make your wall lengths conform to multiples of the length of your standard slipform; 3) Where possible consolidate your plumbing areas and your chimney flues to conserve costs; 4) Plan in alternative energy systems and/or specialized storage areas whether or not you plan to use them immediately; 5) Bear in mind that stone walls in the interior of the house are massively dominating affairs, and that they are sometimes better used in accents rather than dominating themes.

You will find as you progress with your floorplans that other dimensions of your house will begin to encroach upon your thinking Yield to the temptations to interrupt work on your floorplans and sketch these other dimensions while they are still fresh in your mind. Forego artistic endeavor in favor of getting your thoughts down. Before you know it, you will find yourself with a stack of roughly drawn plan views, elevations, sections and details, all of which will stand you in good stead when it comes time to draw your plans to scale — which is the next step.

Scaled drawings do not need to be made into blueprints. Blueprints are for communication between two strangers, the architect and the builder. Since you fill both of these roles you can eliminate the whole business. You can draw your house to scale with no more than a ruler and a pencil, but the whole process is less painful if you invest in an "architect's scale" (a ruler that is graduated in increments or units of twelve) and some corresponding graph paper. The minimum number of scaled drawings that you should make is five: one plan view (a detailed floor plan with one drawing per level of the house), and four compass-oriented elevations (side views).

Elevations constitute another dimension of your house, and as you draw them you will find more problems emerging. If the house you are planning is split-leveled or multi-floored, you will find that roof planes can cause headaches of gargantuan size. Openings in your walls for doors and windows can complicate your floorplans, and, more often than not, entail re-drawing of the latter. But as the challenges are great, so are the rewards. These exterior views allow you to absorb the "feel" of the house. Height and width dimensions that were mere figures before now take on new meanings as the proportions of the house take shape on your drawing board.

Like the Nearings, we designed all of our wall openings to end at the roof plate and thereby avoid stone linteling or arching. This design, combined with our esthetic determination to keep the house's overall profile low, kept the heavier work that we would have to do while forming and pouring the exterior walls to a minimum.

In seeking proportion for our house, we subscribed to Frank Lloyd Wright's idea when he said, ". . . planes parallel to the earth in buildings identify themselves with the ground." We therefore made our house a one-floor structure (with a basement), and went to some lengths to lower our roofline. This entailed lowering the pitch of the roof to one that was appropriate for its probable snowload, and making the exterior wall's vertical dimension end seven feet above the finished floor level. We then beveled our ceilings from this plate to achieve the eight foot ceilings that we sought (see figure 4-1). The latter design, which may be regarded as a deviation from normal design practices, allowed us to lower our roofline one extra foot.

Rooflines should give the builder/designer pause for thought in that the process of drawing elevations reveals how integral they are to the feel of a house. Too often designers are intimidated by accepted modes and styles, and it requires repeated self-reminding to make your house your own.

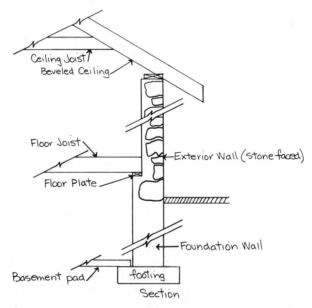

Ceiling Joist
Beveled Ceiling
Floor Joist
Exterior Wall (stone-faced)
Floor Plate
Foundation Wall
Basement pad
footing
Section

FIGURE 4 — 1

FIGURE 4 — 2

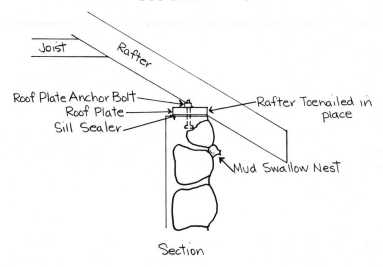

Section

Be guided in drawing your roof by how functionally it will shelter you, its ease of construction, and how you feel it ties with the stone-walled house beneath it. As a general rule, the more facets there are to a roof, the more difficult it is to construct and maintain. Wright, in his inimitable way, put it this way, "Every time a hip or a valley or a dormer window is allowed to ruffle a roof, the life of the building is threatened."

Figure 4-2 shows the detail of how we joined our roof to our stone wall. You will notice that the section drawing includes a mud swallow's nest that was built this past year under the overhang. Like the human occupants, our swallow resident appreciates the snug, sheltered effect of the overhang. Overhangs and overlooks should be designed for your house on the basis of protection they afford from the elements and the amounts of sunlight that they admit or exclude. Prospective swallow residents should take second place in your considerations. In drawing your overhangs or overlooks, you should make a careful analysis of your sunlight needs (see table 3-3 Chapter 3).

Sunlight is a non-variable — that is, you can count on it rising and setting at certain times. Your needs are nowhere nearly so reliable. In drawing plans for your house, you should provide for change. Change could take the form of a physical one, like an addition(s) to your number, or it could be a growing esthetic need for more space.

How do you go about providing for these kinds of possibilities? One way is to design the underpinnings and ceilings in such a way that they require a minimum of supporting walls. In the resultant open space you could, with a minimum of effort, re-define the interior space of your home at a later date. Or, if you feel the future holds even more drastic changes, such as the house's size, you may want to plan one or more facets of the exterior walls in wood or glass. These materials are much more alterable than stone, and if the joining of stone and wood is marked by a vertical line rather than a horizontal one, the results can be pleasing.

Here on our farm we find that our needs are changing daily. Anticipating some of the energy shortage problems, we planned into our house a wood heating system. This was optimal for our location and resources, but had we built in a more southern clime, we would probably have considered other options. Alternative energy systems such as solar, wind, thermal or water are legitimate concerns of the owner/designer-builder today, and there is no better time than when designing your own home to consider them.

There is not enough room in the province of this small book for us to delve into these with the attention they deserve. We therefore recommend that you research these alternatives at some length (see the bibliography) before you really wrap up your plans. Whether or not you actually incorporate one or more of these systems, you would be well advised to provide for their possible later inclusion.

Vegetable gardens, for example, are an increasingly important consideration for many homebuilders. This year will see an estimated eight million gardens established in this country as people return to the practice of growing their own food (in whole or in part), and as a designer you may well want to provide for the processing and storage of this food. At this writing we are, as a case in point, planning an old-fashioned cold cellar. Our idea is to augment if not replace a refrigerator. We were not foresighted enough to have provided for this idea when planning our house, and must now make difficult, and costly, alterations. Fortunately, our alterations will not affect the basic structural engineering.

It is these fundamentals of structural engineering that normally baffle the do-it-yourself designer. Some of these fundamentals were touched on when we discussed siting a house in earthquake areas or areas where frost heaving is a factor. We said that stone walls are compressively strong, but tensilely weak. What did we mean by this?

Most of us, at one time or another has watched with amazement as an expert in the oriental martial arts exhibited his callouses by breaking bricks or cement slabs with his bare hand — some we have seen have even managed this feat with their foreheads. These gentlemen are taking advantage of the basic structural weakness of unreinforced assembled masonry forms. Figure 4-3 illustrates the difference that would occur in this demonstration were the masonry reinforced. When this

FIGURE 4 — 3

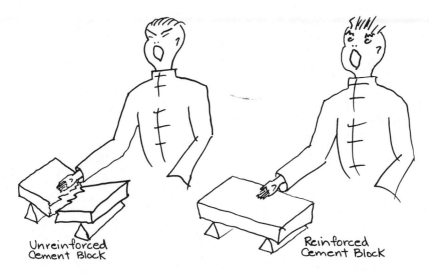

Unreinforced
Cement Block

Reinforced
Cement Block

reinforcement is placed in the masonry nearest the tension side, or the side opposite from that on which the "push" is applied, the "strength" of the masonry is improved.

Figure 4-4 illustrates how we used reinforcing in building our footings, foundations and slab floors. We had erected thirty eight inches of wall using expensive steel reinforcing rods before we realized that at that rate our budget allotment for reinforcement was going to be severely strained. We began scrounging for alternatives. Happily we discovered that a nearby stone quarry was using a steel "sawblade" (heavy-cabled wire) of high tensile strength. The quarry discarded these sawblades the moment they showed wear, and for junk iron prices we acquired enough of them to reinforce the remainder of our entire house, and they now band our entire house at eighteen inch intervals. We added further reinforcement at the corners (one of the weakest points of a stone house) by tying in eight-foot lengths of double point barbed wire — alternating them with the sawblades. This gave us horizontal reinforcement at our house corners every nine inches.

This kind of recycling of societal and industrial wastes augments the challenge of building your own stone house with the perverse satisfactions of "beating the system." Granted these sawblades were not as effective for the reinforcing purpose as was the more expensive commercial reinforcing rod, but we made up in quantity what we lacked in quality. In turn, the use of the cheaper, scrounged materials allowed us the luxury of "overbuilding" the structural underpinnings of our

FIGURE 4 — 4

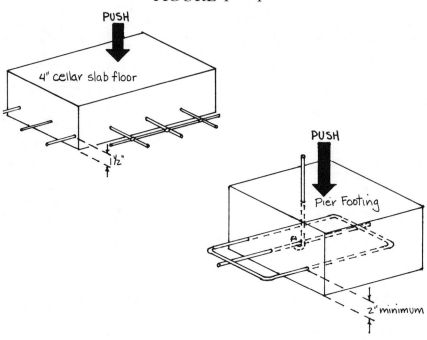

PUSH

4" cellar slab floor

1½"

PUSH

Pier Footing

2" minimum

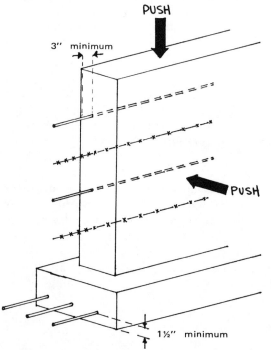

REINFORCEMENT PLACEMENT (Tension Side)

PUSH

3" minimum

PUSH

1½" minimum

house. "Overbuild," like "overkill" is a coined word of expertise. It means to build beyond the accepted norm — a norm that turns out more obsolescent garbage construction each year than our glutted society can assimilate.

We built to suit ourselves, and felt that the additional costs of overbuilding the underpinnings for our stone house were minimal. Our determination to build a sturdy home allowed us the luxury of laughing at some of the "norms", knowing that our structural underpinnings always exceeded the standards set out by FHA and other governmental organizations.

Ground loading is a case in point. We pointedly ignored the maximums allowed for ground loading since we preferred the absolute assurance that our stone house would not settle. As was noted above, stone house walls will take an awful lot of push, but very little pull. "Pull" forces, like those exerted by frost heaving, means that northern builders *must* put their foundation below the frost line.

Settling is a similar "pull" force, and unless the foundation is sited on solid bedrock, it should be footed sufficiently to assure that it will not succumb. This assurance can be had by spreading the weight resting on the footing to a load that is compatible with the site's soils.

Perhaps we should back up a bit to describe what ground loading is, and how it works. One day last summer we were "hunkering" with our neighbor Ken Alger. About sixty of his Ayrshire cows were milling in the barnyard, and we remarked how little his barnyard was torn up despite the crowding. We had just left our pig sty where we kept three small Yorkshire pigs, and the ground there was a chaotic mire. The answer, we agreed, lay in the animals' relative ground loading. After a little mental arithmetic we figured that Ken's 1600 pound Ayrshires loaded the ground under each six inch hoof with about 2,800 pounds per square foot (when standing still). Our 260 poind Yorkshire pig, by contrast, has a two inch hoof that loads the ground with 4,000 pounds per square foot (also when standing still — a rare occasion). The conclusion to be reached is that the weight of an animal (or a house) is immaterial so long as it has a big enough foot.

When drawing up your plans you will want to know how big that footing needs to be. To do this you will need to know your house's: 1) floor area (square feet), 2) roof area (square feet), and 3) perimeter (running feet). You will also need to know the average width and height of your wall from the roof plate clear down to the footing (see how the scaled sections and elevations of your house come in handy?)

The first thing you will need to ascertain is the total weight of your house. Get out your pencil and scratch paper and figure along with us. Don't be shy; a few hundreds of pounds one way or another isn't going to make or break the footing. Assume that the weight of all of the interior of your house (excluding the stone walls), its roof, floors, ceilings, furniture, a box-social crowd of people complete with a blue-grass string band weighs in the neighborhood of 100 pounds per square foot of floor and roof area. This is an exorbitant short-hand figure used by architects and others to calculate the load weight of a conventional frame house, *including the weight of the exterior wall framing and sheathing.* Using our house as an example, our floor area is 1700 square feet, and our roof area is 2500 square feet. 1700 plus 2500 equals 4200 square feet. 4200 times 100 gives us 420,000 pounds of weight, right? We will call this the gut weight of the house. Now put the gut weight aside for a minute, and we will get back to it.

Remember that we have not yet figured in the weight of our stone walls. Look at your elevations and/or sections to determine the average width and height of walls. If you are in a severe frost line area, as we are, you have walls that extend down below ground level for three or four feet. You may have a partial or a full basement — figure it all in. Our average wall width is one foot, our average wall height is 15 feet, and our house's perimeter is 200 feet (look at your own scaled floorplan and add up the facets). In order to arrive at the weight of your wall you must determine the cubic footage, width times height times perimeter. In our case, one times 15 times 200 equals 3000 cubic feet. The average weight of stone and concrete is roughly 150 pounds per cubic foot; therefore, 3000 cubic feet times 150 gives us 450,000 pounds — the weight of our wall. Let's add it all up. The weight of our wall is 450,000 pounds, and the gut weight of the house is 420,000 pounds, for a grand total of 870,000 pounds.

Suppose that you have chosen to build your house on a site composed of sandy soil with a fair amount of clay. Looking at the chart on table 3-1 of Chapter 3, you have already found that the soils will support two or three tons per square foot, but to be on the absolutely safe side you set your load limit at one ton (2000 pounds) per square foot. Using our house as an example the total weight is 870,000 pounds, and we want to find out how wide the footing must be to spread the load out to 2000 pounds per square foot. So, we divide the total weight by the desired loading, 870,000 ÷2000 = 425 square feet. To allot this square footage around the perimeter, we then divide this square footage by the length of the perimeter (our house's perimeter is 200 feet), 435 ÷200 equals 2.175 feet (2 feet 2⅛ inches). This is the width of our footing.

Now that we have gone through the basic arithmetic functions of figuring ground loading, it is probably safe to confirm the old builder's adage that *on proper soils,* the footing for a typical *one-floored structure* should be double the width of the wall, and as deep as the wall is thick.

FIGURE 4 — 5

GROUND LOADING

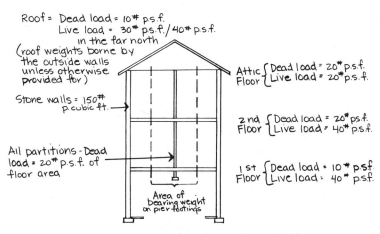

Roof = Dead load = 10# p.s.f.
   Live load = 30# p.s.f./40# p.s.f.
      in the far north
(roof weights borne by
the outside walls
unless otherwise
provided for)

Stone walls = 150#
   p. cubic ft.

All partitions - Dead
load = 20# p.s.f. of
floor area

Attic { Dead load = 20# p.s.f.
Floor { Live load = 20# p.s.f.

2nd { Dead load = 20# p.s.f.
Floor { Live load = 40# p.s.f.

1st { Dead load = 10# p.s.f.
Floor { Live load = 40# p.s.f.

Area of
bearing weight
on pier footings

Dead load = Weight of building materials
Live load = Weight of furniture, snow,
   wind pressure, equipment,
   occupants.

For purposes of calculating the ground loading of the interior underpinnings of your house (pier footings or pigmy walls), you can assume that they will bear the weight of the interior partitions plus one half of all the dead and live weights of the attic, second and first floors. They can bear part of the roof weight if you specifically provide for it with bearing partitions or braces placed under the rafters. Using the dead and live weights shown on Figure 4-5 that calculation for our house looked like this:

Attic floor	½ of 1700 X 40	34,000
Second floor — none		
First floor	½ of 1700 X 50	42,500
		76,500
Partitions	1700 X 20	34,000

Total weight to bear on pier footings   110,500

Since 2000 lbs. was the safe ground loading per square foot that we settled on, we divided the total weight (110,500) by that figure, and thereby arrived at 55¼ square feet. This represented the necessary square footage that the pier footings would have to total. We therefore planned on making ten pier footings, placing them on eight foot centers, and formed them to 3 foot by 3 foot dimensions. This worked out to about 1200 lbs. per squre foot — more overbuilding.

Girders that rest on these pier footings must be selected and sized to suit the load that they will be expected to bear. Table 4-6 shows the standard dimensions required of girders supporting loads for a various lengths of span in one and two-story structures. These figures assume a good quality of wood, and that the girders are solid (not prefabricated). You must, however, use your common sense, and if you anticipate that you will have some extraordinary live weights that you want to impose on these underpinnings you must provide for them. If, for example, you plan on installing a water bed or any other outlandishly heavy equipment, you will have to either, 1) increase the size of the girder, or 2) lessen the span.

Girders are usually employed to support floor joists — ceiling joists more often than not rely on wall partition plates for their support. For purposes of planning, you will need to know how often to place joists, and how long a span they can cover. Table 4-7 assumes a fifty pounds per square foot load (live and dead weights), and a good quality of lumber being used. Using this table, you should be able to determine the placement of supporting girders, and you should, as a result, be able to determine your joist lumber needs.

One of my neighbors is an avid Contra-dancer. Contra-dancing is similar to square dancing and is peculiar to this part of New England. At any rate, it entails a sprightly crew bouncing about with everyone coming down on the floor at the same time. This neighbor tells me that our old town meeting hall is preferred by this dance group "because the floor is easy on the legs." He claims that the joists were placed so far apart when the hall was built that the floor has a lot of "spring" to it, thereby allowing the dancers to carry on into the wee hours before tiring.

We have not dug into this old building to find out whether our neighbor's claims are true or not, and since the building was constructed in 1839 we cannot communicate with the builders, but it all makes sense. All floors have a flex in them. This is sometimes desirable, as in the case of the dancers, but most often it is disagreeable. For general purposes, the center of your span of the floor joists should not flex more than 1/360 of its length — this means a deflection of about ½ inch over a span of fifteen feet.

If you have absorbed the information in this chapter up to this point, you should be well on your way to completing a set of working plans. You should now begin to consider making up your shopping list (bill of materials). Since you are building a stone house, you can expect your list to be weighted with concrete ingredients. To estimate how much sand and gravel you will need, calculate the walls, pier footings, and basement pads (if any) separately. Figuring out how much concrete you will need for the walls is easy once you decide what size stones

TABLE 4 — 6
MAXIMUM SPANS FOR GIRDERS

Girder Dimensions	Clear Span — Maximum	
	one-story house	two-story house
4 x 6	5'	4'
6 x 6	6'	5'2"
4 x 8	6'4"	5'6"
4 x 10	8'	7'
6 x 8	8'	7'
6 x 10	9'	8'

TABLE 4 — 7
MAXIMUM SPANS FOR FLOOR JOISTS

Joist Dimensions	Joist Spacing Center to Center	Maximum Clear Span
2 x 6	12"	10'
2 x 6	16"	9'1"
2 x 6	24"	8'
2 x 8	12"	13'3"
2 x 8	16"	12'1"
2 x 8	24"	10'7"
2 x 10	12"	16'8"
2 x 10	16"	15'3"
2 x 10	24"	13'5"
2 x 12	12"	20'1"
2 x 12	16"	18'5"
2 x 12	24"	16'2"
3 x 8	12"	15'4"
3 x 8	16"	14'1"
3 x 8	24"	12'4"
3 x 10	12"	19'3"
3 x 10	16"	17'8"
3 x 10	24"	15'7"

(average) you plan on using to face the wall with. We have found that our stones occupy a little over half the volumetric space between our forms. That means that our sand and gravel needs are roughly one half the volume of our wall. We make our wall calculations this way. The dimensions used in the formulas should be *in feet*. If you have figured them in inches, the total result must be divided by 1728 to change to cubic feet:

$$\frac{(\text{Width X Height X Perimeter}) \div 27}{2} = \text{Cubic Yards}$$

Remember to include the footing into these calculations.

Pier footings are not normally faced with stone, therefore the formula is simple. If the pier footing is a square or a rectangle:

$$\frac{\text{Width X Length X Height}}{27} \text{ X no. of piers} = \text{Cubic Yards}$$

To find the cubic yards in tapered pier footings, determine the area of the base and top of the pier (width times length), and divide by 2. This will give you an average area. The formula is then:

$$\frac{\text{Average area X Height}}{27} = \text{Cubic Yards}$$

Pads are figured on the formula:

$$\frac{\text{Width X Length X Height}}{27} = \text{Cubic Yards}$$

Totaling the cubic yards for walls, piers and pads will give you your sand and gravel needs. If you are using bank run (sand and gravel that is mixed), this is the figure that you will use for your final estimation. But if you are segregating your sand and gravel, you will need to apportion your estimate to the mix you have chosen. A mix is a proportional "recipe" of cement: sand: gravel. We used a 1: 3: 4 mix, which means that, had we obtained our aggregates separately, we would have needed 3/7 of our cubic yardage in sand, and 4/7 of it in gravel. Similarly, if we had used a 1: 2: 4 mix, the sand portion would be 1/3 of the total aggregate, and the gravel would have been 2/3.

Using this same apportionment, it is easy to arrive at the requisite amount of cement. A 1: 3: 4 mix would require 1 cubic yard of cement for every 7 cubic yards of aggregate. Thus 21 cubic yards of aggregates, using a 1: 3: 4 mix, would require 3 cubic yards of cement (3: 9: 12). Remember, however, that these figures are in cubic yards. Cement usually comes in 1 cubic foot bags, and it takes 27 of them to make a cubic yard. You would therefore need 81 bags of cement.

Framing estimations are the easiest of all to determine. For the shell of your house they include: joists, rafters and partition studs. By far the simplest method to estimate this framing is to draw them onto your

TABLE 4 — 8
SPACING TABLE

Spacing	Constant
12″	1
16″	3/4
18″	2/3
20″	3/5
24″	1/2

plans, and then to count them. If, however, all you want is a rough estimate of the number (the length should be apparent from the mensuration on your scaled drawings), multiply the length of the wall carrying the framing by the spacing constant on table 4-8. This will give you the number of spaces — so you must add one more for a starter. In the case of joists, add one more length for each partition that parallels the joists. For every opening (door, window, etc.) in a partition, add 3 more studs to allow for cripple studs, headers, trimmers, etc. The number of rafters is ascertained by this same method, and extra pieces are included in the estimate to allow for headers and trimmers around chimney openings. Further information that you may need to estimate more complicated rafter needs can be found in Chapter 9.

To estimate how much lumber you will need to sheath your floors (subfloor), you will need to know the square footage involved. Figure 4-9 illustrates how to figure the square footage on a house having offsets or insets. The general rule for calculating area square footage is to

CALCULATING AREA OF FLOORS AND ROOFS

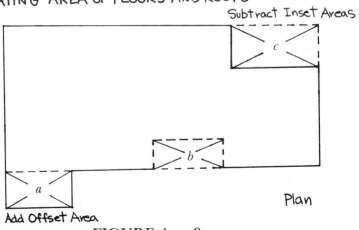

FIGURE 4 — 9

FIGURE 4 — 10

CALCULATING ROOF AREA

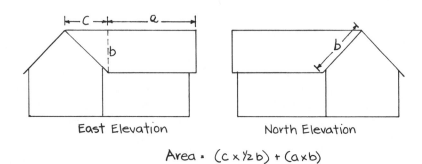

East Elevation   North Elevation

Area = $(c \times \tfrac{1}{2}b) + (a \times b)$

multiply the width dimension by the length dimension. This rule holds true for houses having offsets and insets after you subtract b and c and add a. In your estimations you must add to the net area to account for wastage due to sizing and/or lapping tongue-and-groove (T&G) or shiplap. The rough rule-of-thumb is to add 10 per cent for straight board stock, and 25 per cent for T & G.

Estimating the sheathing for your roof can be a more complicated project. The best way to calculate the area of your roof, and thereby estimate your lumber needs for the sheathing, is to use the measurements from your scaled elevation drawings. Shed roofs present no problems, but gabled roofs with hips and valleys require that you calculate the area of triangles, see Figure 4-10. Figure each facet of your house separately, and then add them together to arrive at the net area of your roof in square feet. Subtract from this net area that square footage taken up by chimneys and insets, and, finally, add 10 per cent to account for the inevitable wastage and sizing.

Bills of materials for lumber are always made up in board feet. To total up your needs in a form that a sawyer or a lumberyard person would understand, refer to the table 4-11 lumber reckoning table on the opposite page. This final bill of materials will include many more items than are included here, but the purpose of helping you to rough out the overall costs you will incur in putting up the shell of your house has, we hope, been accomplished.

TABLE 4 — 11

Board Feet

Nominal size (in.)	Actual length in feet								
	8	10	12	14	16	18	20	22	24
1 x 2	———	1-2/3	2	2-1/3	2-2/3	3	3-1/3	3-2/3	4
1 x 3	———	2-1/2	3	3-1/2	4	4-1/2	5	5-1/2	6
1 x 4	2-2/3	3-1/3	4	4-2/3	5-1/3	6	6-2/3	7-1/3	8
1 x 5	———	4-1/6	5	5-5/6	6-2/3	7-1/2	8-1/3	9-1/6	10
1 x 6	4	5	6	7	8	9	10	11	12
1 x 7	———	5-5/6	7	8-1/6	9-1/3	10-1/2	11-2/3	12-5/6	14
1 x 8	5-1/3	6-2/3	8	9-1/3	10-2/3	12	13-1/3	14-2/3	16
1 x 10	6-2/3	8-1/3	10	11-2/3	13-1/3	15	16-2/3	18-1/3	20
1 x 12	8	10	12	14	16	18	20	22	24
1-1/4 x 4	———	4-1/6	5	5-5/6	6-2/3	7-1/2	8-1/3	9-1/6	10
1-1/4 x 6	———	6-1/4	7-1/2	8-3/4	10	11-1/4	12-1/2	13-3/4	15
1-1/4 x 8	———	8-1/3	10	11-2/3	13-1/3	15	16-2/3	18-1/3	20
1-1/4 x 10	———	10-5/12	12-1/2	14-7/12	16-2/3	18-3/4	20-5/6	22-11/12	25
1-1/4 x 12	———	12-1/2	15	17-1/2	20	22-1/2	25	27-1/2	30
1-1/2 x 4	4	5	6	7	8	9	10	11	12
1-1/2 x 6	6	7-1/2	9	10-1/2	12	13-1/2	15	16-1/2	18
1-1/2 x 8	8	10	12	14	16	18	20	22	24
1-1/2 x 10	10	12-1/2	15	17-1/2	20	22-1/2	25	27-1/2	30
1-1/2 x 12	12	15	18	21	24	27	30	33	36
2 x 4	5-1/3	6-2/3	8	9-1/3	10-1/3	12	13-1/3	14-2/3	16
2 x 6	8	10	12	14	16	18	20	22	24
2 x 8	10-2/3	13-1/3	16	18-2/3	21-1/3	24	26-2/3	29-1/3	32
2 x 10	13-1/3	16-2/3	20	23-1/3	26-2/3	30	33-1/3	36-2/3	40
2 x 12	16	20	24	28	32	36	40	44	48
3 x 6	12	15	18	21	24	27	30	33	36
3 x 8	16	20	24	28	32	36	40	44	48
3 x 10	20	25	30	35	40	45	50	55	60
3 x 12	24	30	36	42	48	54	60	66	72
4 x 4	10-2/3	13-1/3	16	18-2/3	21-1/3	24	26-2/3	29-1/3	32
4 x 6	16	20	24	28	32	36	40	44	48
4 x 8	21-1/3	26-2/3	32	37-1/3	42-2/3	48	53-1/3	58-2/3	64
4 x 10	26-2/3	33-1/3	40	46-2/3	53-1/3	60	66-2/3	73-1/3	80
4 x 12	32	40	48	56	64	72	80	88	96

Abstracted from: *Carpentry and Building Construction,* Department of the Army Technical Manual 5-460.

"Structure is simply function — in slow time."

Bertallanfy

The authors-builders, Sue and Karl Schwenke, pose briefly on the house site after breaking ground. RICHARD WIEBE

The Schwenkes Build Their Stone House

The footing trenches and partial basement have been excavated and the batterboards are up. Footing forms come next.

The Schwenke's latest stone building, a barn, was started in late August and was finished before the first snowstorm December 1st.

The slipforms have been wired, plumbed and braced; now are ready to stone and pour. Uncredited photographs are by Karl Schwenke.

Ascending the walls in leapfrog fashion, the forms here are nearing the plate level. The form on top is all ready for laying the stones. Note inset basement window below.

This is how the house project stood after four months of steady work. At times it seemed to go at

Sue took care of the whole untidy operation of mixing concrete for the forms.

ail's pace and then there would be a surge as separate projects came together.

The outside wall now is completed to the plate, but work continues on the gable ends.

The final backfill and landscaping went on even as they were erecting the rafters. They were hurrying now, since the leaves were falling and each morning they found ice on the water barrel.

Here is a section of the finished house. There wasn't time to complete the chimney exterior until the next spring.

The fireplace gave the Schwenkes the new experience of hand-laying the stone. This was done around metal heat-exchanger units built from a cantilevered hearth.

Examples of Other Stone Houses

This gambrel-roofed slipform house was built by the Schwenke's neighbors, with wood used above the plate. Stone work on the left side has been pointed with trowel and whisk broom. Right side is as it emerged from the slipforms.

"The Castle", a well-kept Victorian
showplace built in 1893, employed
irregular, mortared fieldstone laid to the
second floor level and for the pillars and
chimneys.

Not far from the Schwenke's slipformed
home is this snug cottage, built about 150
years ago by Scottish settlers from dressed
stone quarried nearby.

This ornate "rope" mortaring technique was used in building The Castle (left). Probably lampblack was mixed with the mortar to achieve the color, and special trowels cut from pipe sections were employed.

Tools & Slipforming

Chapter 5

The brainwork of the drawing board ended none too soon for us. For while we had enjoyed the challenges of researching, siting, planning and estimating, we were primed to get underway with the real business of building. Preparatory to starting the house, we had accumulated a lot of tools, and with the warming weather of spring, we itched to put them to work.

One old-timer whom we consulted told us that all we needed to own by way of tools was a hammer, saw, shovel, trowel, plumb bob and a framing square. "Y'won't need nothin' else," he said "'less'n its ree-placement thumbs when y'mayush the ones yer wearin'." This old-timer could probably have done the job with these personal tools, but it is our experience that before he was done, he would have worn a path between his building site and his neighbor's tool shed.

Here is a list of tools we used every day:

1 Cement Mixer	Whether powered or hand-driven, it should be capable of mixing one-third bag mixes. Bought new, powered mixers usually come with one-third horsepower motors. We found this inadequate, and would suggest that the smaller motor be exchanged for one of at least one-half horsepower.
1 Wheel-barrow	The wheelbarrow should have a five cubic foot capacity, and the higher sided "construction" type is preferable to the shallower garden variety.
2 Shovels	Like the hammer, the shovel is an intimate tool. Care should be exercised to select one of good steel, and it should be handled of a good, straight-grained ash or hickory.
1 Pick	A skillful pick man is a joy to behold at work, and we recommend that the do-it-yourself house builder, par-

ticularly if he is undertaking to do his own excavation work, get acquainted with this important tool. In choosing a pick the length and weight of the head is critical. If you intend the pick to be used by a person of a lighter frame, you should choose one with a head weighing six to eight pounds, and measuring twenty to twenty four inches across the length of the head.

1 Hoe   A masonry-type hoe (large blade with holes) is a good choice if you anticipate a great deal of hand mixing, but the large-bladed garden type is otherwise adequate, and it is versatile.

2 Hammers   Of the various claws available on most hammers, we found the rip-claw (straight claw) to be the most useful, particularly for form work. At present, hammer handles are made of wood, metal and fiberglass. We have used all three, and agree that metal handles prove more durable and trouble-free. When selecting your hammer be sure that the head weight is at least sixteen ounces — more if you can handle it.

2 Handsaws or   Handsaw needs will require you to have one with
1 Electric   crosscut teeth, and one with rip teeth; (there is nothing
Handsaw   quite so miserable as trying to cut the length of a wet board with a crosscut saw). Metal in handsaws must be of top quality, and the blades should be kink-free. Electric handsaws (more commonly called "Skill-saws") are a different kettle of fish. Choosing one is often a matter of electrical aptitude. Suffice it to say, the general rule-of-thumb is to choose one of the higher horsepower. These saws should be fitted with a "rough-cut, combination" blade for all-round construction work. Other blades for specialty jobs will have to be purchased.

2 Sixteen-foot   Sixteen-foot tapes are bulkier than the shorter ones,
Retractible   but their wider blades allow them to be manipulated
Measuring   more readily when extended, and the rigidity is indis-
Tapes   pensible when measuring vertical distances. We found this tool our most commonly-used item in construction.

2 Hammer-   These are metal rings sewn into leather patches. The
rings   leather is slitted or doubled to accept a belt, thereby providing accessibility and portability.

It should be noted that the above list of tools is the product of the two of us working together. The following list is that of our lesser-used, but also everyday tools:

1 framing square
1 thirty-inch wrecking bar
1 brace with assorted bits or
1 electric drill with assorted bits (½″ or larger)
1 set of chisels (⅛″ — ¾″)
1 jackplane
1 hacksaw (with extra blades)
1 pair of tinsnips
1 pair of diagonals
1 pair of pliers
1 eight-inch crescent wrench
1 plumb bob with string
1 masonry or carpenter's level
1 100′ chalkline with chalk
1 standard, large screwdriver
1 standard, small screwdriver
1 phillips-head medium screwdriver
1 100′ measuring tape (steel or cloth)
1 axe (¾ length or larger)
1 finishing trowel
1 large, pointed masonry trowel
1 small, pointed masonry trowel
1 bastard file, large
1 bench stone (8″ or larger)
1 grease gun and grease
1 pair heavy rubber gloves

The following is a list of tools we have that would not be absolutely necessary for the beginning builder to acquire. You might call them optional, but nice to have.

1 electric saber saw
1 set of masonry-tipped drill bits
1 builder's level (sometimes called a "dumpy level")
1 bench vise with six inch jaws
1 sledgehammer, eight pound
1 bolt cutters
1 pry-bar, five foot
1 bench grinder

Clearly not everyone has all of these tools in his possession before building, and the initial purchase price of tools on these lists is quite high. Bought new the cost today of these tools might come to around

$600. But, as was true in our case, most people who do a modicum of do-it-yourself household repair have a portion (sometimes most) of the tools listed here. Even though the tools one may already have are not of the first quality, they can be made to "make do" until the opportunity and/or the finances allow you to acquire them.

The alternative to buying your tools new is to buy them used. Used tools can be picked up cheaply from a variety of sources. Some of the more memorable moments before and during the early stages of building our house were those precious hours we took out to attend tool auctions, rummage through pawn shops, visit garage sales, and haunt salvage yards. We acquired our total list of tools (including the "optional, but nice" ones) for less than $300, and we enjoyed ourselves in the process. This approach, however, requires advance time, and fair-sized dollops of patience. You have to know what you are looking for, how much you should pay for it, and, above all, you must know quality when you see it.

Quality in hand tools is becoming increasingly difficult to come by, and more often than not, you have to pay extra for it. To stint in the purchase price of a tool is penny wise and pound foolish. A good tool, with adequate care, should last long beyond your present building project, and each passing year will depreciate the high initial cost of almost any hand tool to pennies before it is worn out. But wear out they will, despite all the loving care you may lavish upon them. Tools that you use most often begin, after a while, to take on the character of old friends.

Sue did most of the mixing of the concrete for our stone house, and she soon developed a preference for one of the two shovels we had. It was the "good feel" of the handle in her hand, she said. If I mistakenly picked up "her" shovel, she insisted on swapping, and she showed her attachment for this tool by caring for it faithfully. I watched the nose of that shovel wear down until it was flat across the end. In the ensuing weeks, I deliberately refrained from comment as I watched the flattened end continue to erode inward until there were about four inches gone from the original shape. One hot summer's afternoon, after we had finished a particularly difficult pour, I came upon Sue digging a long hole next to the footing of the house.

"What are you up to?" I asked.

"Digging a grave," she responded solemnly. Her face was cast in appropriately funereal lines. At that point she finished the shallow "grave," and with no further words she threw her trusty shovel into the hole and covered it over.

Among the many lessons we learned from Scott and Helen Nearing was the invaluable work of caring for tools. Having a place to

FIGURE 5 — 1

CARPENTER'S NAIL APRON

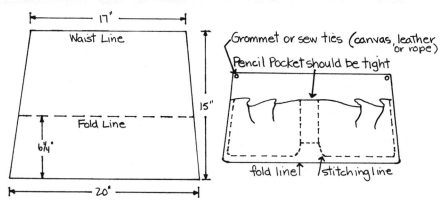

store them is the first order of business in building a stone house. Once started, it did not take us long to build a tool shed. We made it of scraps, and it more than justified the time it took to put it together. Like the Nearings, we made it a standard, daily practice to return the tools we used each day to the tool shed, and, thanks to the Nearings again, we wiped each tool with used motor oil before putting it away. In our tool shed we kept a twenty-gallon can filled with the motor oil emptyings from our vehicles, and it was a simple process to dip each tool before storing it. Additionally, we purchased a small, cheap hand grease gun with which we gave a daily dose of grease to wheelbarrows and cement mixers. (We also use it to service our own vehicles.)

But hand tool care does not end with this. All edged tools should be carefully sharpened as the need for it arises — this includes shovels, axes, planes, chisels, bits and saw blades. Other tools like screwdrivers, trowels, picks and hammers need occasional dressing of their working ends. After sharpening or dressing, the tool is not only easier to work with, but it will last longer, and it is far *safer*. Nearly all sharpening and dressing chores can be accomplished with a large bastard file and a good bench stone. With these two sharpening and dressing tools, it is next to impossible to ruin the temper of a prized tool. It is this fact that persuades many craftsmen to forego the easier-to-use electric bench grinder.

Other tools you will need to build a stone house are ones that you can make yourself. They include: carpenter's aprons, sawhorses, a mixing trough, a sand and gravel sifting screen, and slipforms. Figure 5-1 shows about the simplest carpenter's apron that you can make. The measurements taken across the apron can vary, depending upon your lap. A friend, who cheerfully admits to being "about four axe handles"

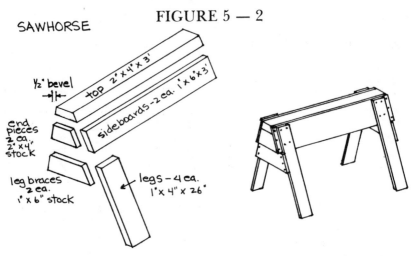

SAWHORSE

FIGURE 5 — 2

½" bevel

top 2"x4"x3'

sideboards-2 ea. 1"x6"x3'

end pieces 2 ea. 2"x4" stock

leg braces 2 ea. 1"x6" stock

legs-4 ea. 1"x4"x 26"

wide across the beam, designed his apron with an accommodating four-pocket width.

Sawhorses are indispensable tools for house building. They are best constructed in pairs, as this is the fashion in which they are generally used. Around the building site their uses are unending. They are used to hold lengthy boards for sawing; they serve as legs for a temporary workbench; they hold work for planing; and the list goes on. Figure 5-2 shows one simple sawhorse design that has proven sturdy and a welcome addition to our tool inventory.

The stone house builder who undertakes to mix his own concrete without a mixer needs to build a mixing trough. Figure 5-3 shows one easily-built mortar trough that incorporates a minimum of materials, and which should prove sturdy enough to survive the rough usage that it will be subjected to. The dimensions of this trough are, of course, variable (depending upon your needs), but you should keep the fact in the back of your mind that you will eventually have to move the trough around. No matter how faithfully you clean a mixing trough, it somehow ends up weighing twice as much after use as it did when you built it. One stone house building site that I visited centered around a huge mixing trough. Mixing their concrete entirely by hand, this young couple were obviously tired by their labors at the end of the day, for the exterior of the trough showed little evidence of any cleaning efforts. Hardened concrete had long since piled up around the trough. "How are you going to move this thing when you're through?" I asked. "Move it hell!" the man responded cheerfully. "When we get done, we're going to fill it with water and put goldfish in it."

A screen is a simple tool to make, and it is indispensable if, as we did, you are using "bank run" sand and gravel. Fireplace building and other forms of handlaying require sifted sand, and projects requiring a washed pebble exterior are coated with the gleanings from this screen. Figure 5-4 shows the basic screen that we made, and, it should be noted, the size mesh indicated in this figure is the one we found to be the most useful.

FIGURE 5 — 3

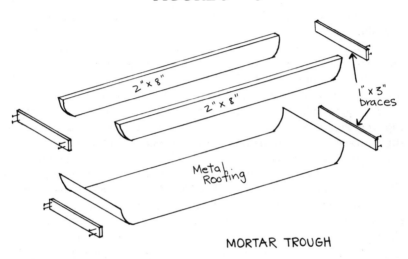

MORTAR TROUGH

FIGURE 5 — 4

SAND AND GRAVEL SCREEN

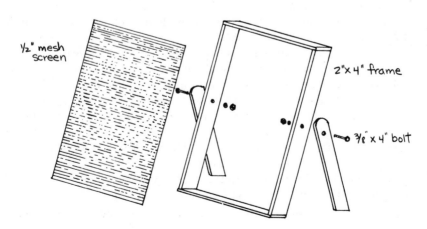

The most commonly-used tool you will make is the slipform. The step-by-step directions which follow will show you how this form got its name. To make a slipform study the details of figure 5-5. This figure shows the appropriate nomenclature of the form and the mensuration for its construction. In the course of forming your house, you will find the need will arise for forms of varying lengths. This need most commonly occurs when forming the interior walls to a corner. Careful pre-planning can often eliminate the numbers of these "odd-ball" forms, but it is sometimes an irremediable fact that interior wall measurements in multiples of eight feet are just not always practical or desirable.

Our "standard" slipform measures eight feet in length, and is nineteen inches high. The measurements are arbitrary — you could make the form longer and higher, if you felt that you could cope with the consequences of increased weight and difficulties of working with it. We chose this particular vertical height because it made a comfortable working depth into which stones could be laid. After some use, our slipform acquired enough moisture and stubbornly adhering concrete to make it a difficult job for one person to lift it to any great height. Our intent at the outset of our building project was to make the form units of manageable size for *both of us*. The 8' X 19" form proved to be the limit in weight and cumbersomeness for Sue. Hindsight indicates that we might well have cut the overall weight of the form with the use of 2" X 3" whalers and form studs — in place of the 2" X 4" stock that we did use (figure 5-5). In a couple of trial forms we found the smaller form stock to be of sufficient strength, but it stands to reason that with continual use, the heavier stock will last longer.

This brings up the question of building projections — that is, just how many buildings or structures do you intend to erect using the slipform methods? As of this writing, our original slipforms are being used to build their fifth structure, and with an occasional repair job it looks as though we will be able to use them for another five. For us durability in our slipform is important, and we would urge prospective builders to try to anticipate beyond their immediate building project to the possiblity that they too may want to build other structures.

The lumber stock cited in figure 5-5 is of the "native" variety. We cut spruce trees for this lumber from our own woodlot, and then had a local sawyer cut the required stock we felt we would need. As a result, the measurements as given are *actual* ones. The lumber order we placed with our sawyer included the 1" X 6" facing boards. We had the latter cut in sixteen foot lengths (good sawyers always give you a few inches of extra length), and sawed them in half when constructing the forms.

In making forms, we first cut all the whalers and form studs. This

FIGURE 5 — 5

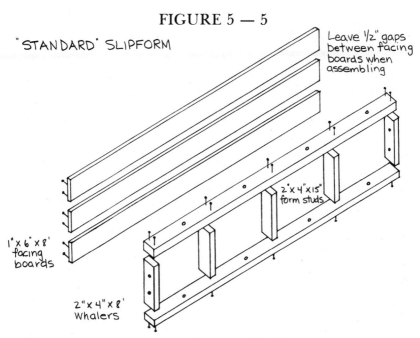

"STANDARD" SLIPFORM

Leave ½" gaps between facing boards when assembling

2"x 4"x15" form studs

1"x 6"x 8' facing boards

2"x 4"x 8' Whalers

was a "production line" job with one person doing all the mensuration and squaring, and the other doing all the cutting. Then choosing one each of the whalers and form studs, we made bolt-hole template patterns. When picking out a whaler and a form stud for the template patterns, be careful to choose for straightness and evenness of grain. Then using the template patterns as guides, we bored three holes in each whaler, and two holes in one half of all the form studs (only two out of every four form studs are end ones). The bolt holes on the whalers are spaced every 24", and on the form studs they are located 4" from the ends of each stud.

Assembling the slipforms should be done in two stages: First, butt nail the whalers to the form studs (use 20d nails with full dimension lumber, 16d nails with commercial lumber). This assembly-line process is monotonous, but we found it the opportune time for Sue to learn the finer points of hammering nails. Second, nail the facing boards onto the assembled frame of whalers and form studs. We originally used 6d nails for this purpose, but found that with extensive use, they tended to pull loose, and we have subsequently used nothing but 8d nails. If, as suggested above, you have your 1" boards cut to 16' lengths, it is faster to nail the full lengths to the frame first, and then cut them off afterwards. Lastly, it makes wiring the forms a simpler job if you leave a small crack between the facing boards (¼" to ½"). This crack does not allow any appreciable amount of concrete to seep out, and it allows for expansion when wet.

SLIPFORMING

The "slip" of slipforming derives from the practice of "slipping" forms up the wall as you erect it (see figure 5-6). The slipform has several good talking points — especially for erecting stone walls. Reusing the form several times as it makes its way up the wall conserves expensive form wood, and its nineteen-inch height makes for a comfortable working depth in which to lay stones. Further economy derives from the fact that in building your own forms, you build in provisions for bracing and wiring.

Theoretically, you could build a slipformed stone house with as few as eight standard forms (four to each side of a corner), but the process would be an interminably long one, and would result in radically different drying times within the walls. This is particularly true if the house you envision is a large one. The number of forms you will need is directly proportional to the size of structure that you will wish to build. While eight forms might well be enough for small structures like tool, potting or wood sheds, the odds are that you will need far more forms for your house.

The actual process of setting up forms, preparing them for a pour

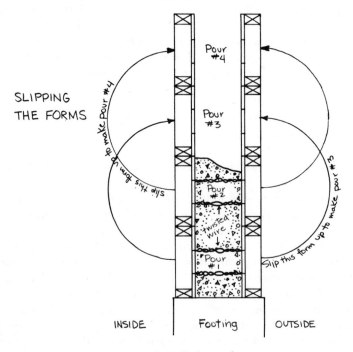

FIGURE 5 — 6

and then stoning them is, in practice, a set of routines that makes the work flow along at a good, steady pace. Once underway, our exterior walls rose at the daily rate of nineteen inches in height by twenty-four to thirty-five feet in length (depending upon weather, breakdowns and other unforseen happenings). Our daily routine was to get three or four standard length forms ready for pouring *in the morning*. This practice left us the entire afternoon to cope with the actual placement of stone and concrete.

Each morning began with what we came to call the "unveiling." This was when we removed the forms from concrete and stone that had dried for a minimum of three days. The stone which we had so laboriously laid in place three days before emerged to view for the first time, and the patterns of color, the shapes and the textures thus revealed were a constant source of matutinal satisfaction — a fine way to start a day.

Removing the forms from a previous pour is not always as easily done as it is said. This is particularly true when it comes to the forms that are first laid on the footing. Here you contend with an aggravating lack of elbow room in that you are working at the bottom of a narrow excavation where "prying" room is at a premium. Further aggravation is added as the friction of the form on top of the one you want to remove defies heavenly imprecations, brawny muscles and an unhealthily large "cussin' vocabulary."

The first time I encountered this problem I had exhausted my imprecations and muscles, and was well launched into my last refuge when I became aware of an onlooker. It was my neighbor, Ken Alger, who stood in slack-jawed, admiring awe of my tantrum. "That's the first time," he grinned "I ever heard a feller carry on like that for so long, and never use the same word twice!"

Unfailingly a good neighbor, Ken always offers more than a good anecdote. Like all New England Yankees, he has an uncanny ability to problem solve. One look at my problem, and he suggested that we use a wedge-shaped shim under future forms that were to be placed on top of the footing (see figure 5-7). We did, and it works fine.

Trial and error led us to several "tried and true" routines in laying up the forms for the afternoon's pour. The first of these routines was to *place* the forms. In the case of putting the first forms on the footing, it meant first chalklining the wall's outline on the footing, and then making the forms toe the line. When lifting forms from one level to the next as in figure 5-6, it is a simple matter of adjusting them so that their respective bolt holes line up, and inserting the bolts. Up to this point you pay no heed to plumbness or bracing, and concentrate on merely getting the forms in place.

The second routine for us was to brace, space and lace (wire) the forms together. The first bracing to be done is across the tops of the forms — so that they are joined together at the proper wall-width distance. This bracing in combination with the bolts joining the forms gives adequate rigidity to allow the wiring process to proceed without upsetting the form placement.

Vertical form junctures were almost always wired together, but in any case, we made it a rule to have at least two wires (23 gauge or larger) supporting each form. Lengths of wire measuring three to four feet (depending on the width of the wall) were cut and draped over the forms. Then, working as a team (one on each side of the wall) we doubled the wire, inserted it between the facing boards of the forms, and guided it through the opposite form. If you are fortunate enough to have form studs that "occur" at the convenient point where your wire is located, it is a simple matter of twisting the wire around the stud. But if, and this is most often the case, you find no handy stud to which you can attach the wire, you must use a bent nail (see figure 5-8).

The nail must be bent because a straight one, when the wire is tautened, digs into the back of the form. This will cause you to chisel around the head of the nail in order to pull the nail and release the form. Bent nails are not hard to find around a construction site — particularly if you have a novice carpenter or two around. Gathering bent nails is an ideal job for young children, or for visiting adults who insist on "being helpful" when there is no way that you can fit them into the work you are doing.

Next, we inserted a nail or a leg of a pliers between the wires, and twisted the wires until they were tight (those with a good ear will recognize a B-flat above middle C when the properly tightened wire is plucked). We found that the twisted wire tended to pull the forms inward unless we used a "spacer", a 1″ X 1″ scrap of wood cut to the width of the wall. Inserting this spacer between the forms, we only had to tighten the wire until the forms contacted the spacer (see figure 5-8) You can make a fine comedy sequence out of this process by tightening the wire too tightly, thereby binding the spacer irrevocably into place.

After wiring the forms, we plumbed and "set" the corners. Setting the corners consisted of placing a level against the outside of the forms, pushing the forms until they were vertically plumb, and then nailing a brace across the top of the forms that join two legs of a corner (both inside forms and outside forms). The final plumbing was done with external braces, usually long two-by-fours. These two-by-fours were nailed to stakes driven into the adjoining excavation, or they were attached to other, secure structures like other walls, floors, etc.. As few as two or three of these external braces were sufficient to hold long stretches

FIGURE 5 — 7

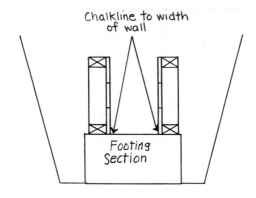

Chalkline to width of wall

Footing Section

FIGURE 5 — 8

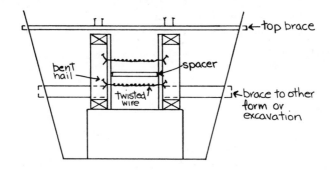

top brace

bent nail

spacer

twisted wire

brace to other form or excavation

FIGURE 5 — 9

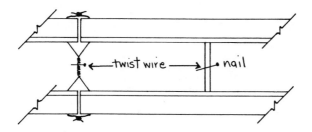

←—twist wire—→ nail

of forms (24' to 32'), provided the forms were properly bolted and braced.

Like the process of hand-laying, the process of slipforming *can* produce a crooked wall. A good stonemason laying up one stone at a time can make a small mistake, like setting in a stone that protrudes a half inch or so, but he easily corrects the error in the next one or two stones that he puts into the wall. But the slipform builder who makes an error of this sort will find the error harder to correct. He must wait until the next level of forms to correct his mistake. Careful use of batterboards (discussed in Chapter 6) will help to get your wall started out right.

Once your forms rise above the level of the batterboards, however, it is a wise policy to check each level of forms with a stretched chalkline. This chalkline, when secured at the terminal corners of a common wall, will show "swaybacking", and "bowing", and it will allow you to correct the fault before you monumentalize it in concrete. Some slipform builders I have observed have been able to erect a faultless wall with a two-foot carpenter's level, and a good "eye," but they were exceptions to the rule.

Our last two routines to prepare for the afternoon's pour were to lay in the reinforcing steel, and to sort out the stones in such a way that they would be readily available to the person who placed them in their final places between the forms. These two routines are described in detail in Chapter 6 and Chapter 7, respectively.

> *"Work is love made visible. And if you cannot work with love but only with distaste, it is better that you should leave your work and sit at the gate of the temple and take alms of those who work with joy."*
>
> Kahlil Gibran

* * * *

> *"There is no jesting with edged tools."*
>
> Francis Beaumont

Footings to Backfill

Chapter 6

B efore excavating you may need to clear your site. There may be underbrush or even trees to be removed. Like most aspects of house building, you must look to the eventual completion of the structure to decide how best to go about the process. Trees are, for example, a long time in replacing, and you had best take a long look into your crystal ball before you begin laying about with an axe or chainsaw. Should you decide that you must cut some trees, it is a good idea to do a workmanlike job of it, and cut it up into fireplace-length cordage. If you cannot use it in your own fireplace, you might well sell it to help defray the costs of excavation.

The next step preparatory to excavation is to lay out your house. We suggest that you do your layout in two stages: the rough layout, for purposes of excavation, and the final implantation of batterboards, which you will use for the erection of the footings and the walls. We found that putting the batterboards up too early resulted in their being knocked about by the heavy equipment that we employed to do our excavation.

Subsequent building efforts showed us that the additional step was worthwhile, and we commend it to you. If, however, you plan on excavating your house's footings and basement by hand, you should ignore the rough layout and set up your batterboards immediately. Hand excavation takes a long time to do — particularly if you plan a basement under your house. If your building season is limited, we suggest that you give thought to employing a backhoe or other power equipment.

The rough layout of your house begins with a close scrutiny of your plan drawings, and a review of your research into building codes that pertain to your area. If these codes require a setback (a specified distance from the road or property line), that is where you begin. Measure this distance, and then mark the imaginary line with a couple of stakes. Next, orient the house making sure that you do not transgress the setback line. Those of independent bent may not wish to follow the conventional idea of having their house front parallel the road, and may opt instead to orient to the sun or a favorite view.

It is a good idea at this point to know whether or not you intend to excavate with power equipment. Many equipment operators prefer that the layout stakes mark the middle of the wall of the house, while others demand that you mark the outer limits of the actual excavation.

Before putting a stake in the ground, you should arrange for the excavation, and inquire after the operator's preferences. For purposes of this rough layout, we will assume that the operator prefers that your stakes mark the outer limits of the excavation. We will further assume, for purposes of illustration, that the house is a conventional rectangle.

Figure 6-1 shows a house oriented to the sun. The first layout steps you should take should be to establish the four corners of the house (as shown on your plan drawings). From these corners you can then run parallel lines to mark the outer limits of the excavation. Begin your layout by establishing one corner with a stake (A), then, following your orientation line, place stake B by measuring the length of facet AB to correspond to the outside mensuration shown on your drawings.

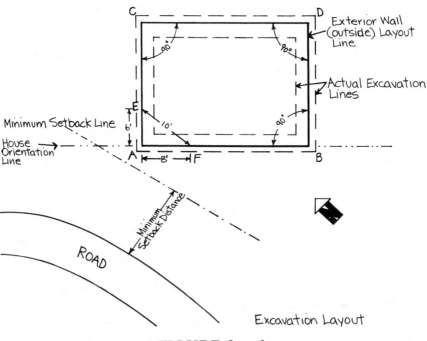

FIGURE 6 — 1

Next, starting from these two base stakes, measure off the width of the house to establish points C and D with two more stakes. Take it easy when pounding these stakes into the ground, as you will most likely have to move them around and assure yourself of 90 degree interior angles. If you measure CD and make it agree with AB, you are to be congratulated on making a nice parallelagram. A professional oddsmaker would give you a hundred to one against placing the stakes perfectly right (so that they made at least two 90 degree angles) the first time that you tried.

To assure yourself that you do have square corners, take your tape and measure eight feet from stake A along line AB. Using your plumb bob, set a stake at this point (E). Then along line AC, measure off six feet from stake A — here you drive stake F. Now measure distance EF. If this distance is ten feet, you have a 90 degree angle and you are well along the road to proving the oddsmakers wrong.

But if, as is most often the case, this measurement does not jibe, you must move lines AC and AB until this distance is right. You won't win any bets until you check at least one more corner, and if it were my house, I would check all four. As a final clincher, stretch your tape out diagonally from corner to corner (CB and ED). If the measurements you get are equal, your house is square. Remember that these stakes mark the outside wall of your house, and you must provide for enough room in the excavation for the footing and working room. We found expanding the measurements of our rough layout by one foot on all sides gave us adequate room for working on the outer edge of the footing. We then instructed the equipment operator to dig the footing excavations three feet wide, beginning from this outer line.

Layout also includes those portions within the house where you plan your basement or fireplace. These layouts are only a matter of careful mensuration. If your original layout was square, your subsequent fireplace and basement layouts will be equally square. A checklist of other excavation work that you might want to lay out could include: sewage lines, septic tanks, drainage fields, surface wells, water lines, incoming underground electrical power wires, telephone lines, solar heat-gathering fields, methane displacement tanks, furnace fuel oil storage tanks, or even footing holes for a tower to hold a wind-powered generator.

Most equipment operators know their business, and your part during the actual excavation will be that of sidewalk superintendent. If you have a backhoe do your job, as we did, the operator will want to know if you want your site "stripped," and how deep the footing and/or basement excavations are to be.

Stripping is the taking off of the first foot of topsoil in order that the site be completely clear of all vegetative material. This practice, which is sometimes codified, is a good idea, as the topsoil would otherwise be wasted. Have the operator set it aside, for it will come in handy when it comes time to do your finish landscaping.

In choosing a backhoe operator, we took the advice of our neighbors, who described this man as being able to "thread a bran sack needle with his backhoe bucket." He was good, and he put on quite a show — a show that began with the curious rite of planting what looked like a Sioux war lance right in the middle of the house site. We contained our curiosity about this artifact-looking object until he had finished the whole job, and then we asked him about it.

"Well sah!" he responded with a typical Vermonter opener, "that's my ree-minder."

·"Your reminder?" prompted Karl.

"Eyuh. Fust cellah hole I evah dug I planted m'self in the middle and commenced in t'diggin'. Wasn't payin- no nevermind, y'know . . . jest a diggin' away. Well mistah! 'Fore I knew nawthin' me 'n ol' Betsy was sittin' on top of the purtiest island you ever did see."

"I see," said Karl laughing. "You put that thing out there to remind yourself not to dig yourself into a corner . . . Say, how did you finally get yourself off that island?"

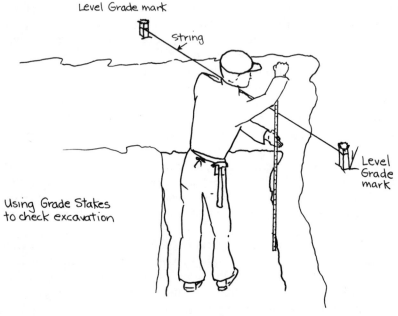

FIGURE 6 — 2

FIGURE 6 — 3

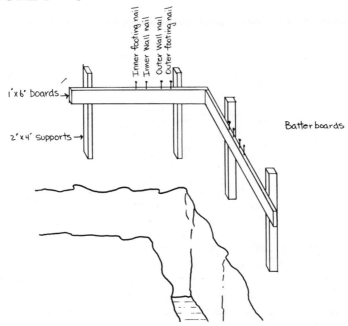

"Well sah! I jest reached over with m'bucket," the operator extended his arm and hand in visual aid to his story, "and got me a good bit of ledge. T'warn't much, but jest 'nough so as t'heist me and ol' Betsy up in the air."

"Sure," said Karl, "You just lifted the entire machine up in the air, balanced yourself, and . . ."

"That's right mistah," he interrupted. "Swung m'self and ol' Betsy right over to terry firma. Y'never tol' me you knew nawthin' about runnin' a backhoe."

The operator never once cracked a smile during the telling of his yarn, and neither did Karl as he wrote out the man's check — for a sum that included the time it took the operator to recount the saga.

No matter how skillful the equipment operator may be, there always remains some hand work to do. You can eliminate a lot of this pick and shovel work by having been around while the excavation work was being done, and having periodically checked the depth of the excavations with grade stakes (see figure 6-2) or with a transit.

With the heavy machinery gone from the site, we next undertook to erect batterboards at the corners of our excavation. Figure 6-3 shows one of these batterboards, and it illustrates how we placed the nails to indicate the layout lines.

FIGURE 6 — 4

FOOTING

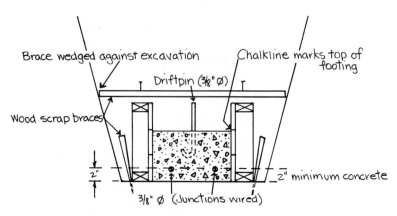

Brace wedged against excavation   Chalkline marks top of footing

Driftpin (3/8" Ø)

Wood scrap braces

2"   2" minimum concrete

3/8" Ø (Junctions wired)

Using the outer wall dimensions, which we took from our scaled plan views, we laid the house out in the same way that we did for the excavation work. Once the outer corners of our wall were established on our batterboards (marked by a nail in the top edge of the board), we transferred the other attendant measurements from our plans to these batterboards.

These measurements included one nail each for the inner edge of the footing, the inner edge of the wall, the outer edge of the wall, and the outer edge of the footing. Running strings from one batterboard to another provided us with a reliable line from which (using a plumb bob) we placed our forms.

With the batterboards in place, we began the building process. Forming the footing was the first order of business, and with all our advance preparation it went quickly. Figure 6-4 shows how we braced these forms using scrap wood. Lengths of reinforcing rods were then placed between the forms, and we then wired the junctions. This wiring saw the rods overlapping one another by at least 24 diameters of the rod. To keep the rods up off the dirt, so that they would have at least two inches of concrete beneath them, we wedged them up with assorted stones. Next, we cut 18 inch lengths of reinforcing rod. These drift pins we bent into a "J" shape (see figure 6-4), and then laid them aside until we were ready to begin the pour.

The final step before pouring was to snap a chalkline along the interior of the forms at the level height of the top of the footing. We found that the small amount of time that it took to float the concrete to the level of this chalkline paid dividends by making it easy to place the first level of forms for the foundation wall.

We used a builder's level to determine this level, but one could use a straight-edge and an ordinary mason's or carpenter's level for the shorter distances. Longer distances, however, require a sighting level, or a level fashioned from a hose filled with water. Both of these methods are home-made, but they are accurate (see figure 6-5).

We were now ready to pour the footing, but as a last gesture of preparation, we stacked stones (for filler) alongside the forms. These stones were the more unattractive or badly faceted rejects taken while sorting our facing stones. Used as filler, they replaced enough concrete to warrant the investment of time required to sort them out and to clean them.

A newly excavated building site is most often a hilly disaster area when it comes to wheeling a wheelbarrow around. We found that the easiest place to run the wheelbarrowsful of concrete was directly across the tops of the forms. To make a firm runway, we inverted several of our standard forms so that the boarded side was up, and laid these between spans of scrap two-inch stock — this stock was placed athwart the footing forms. By ramping to these runways, we found that we had solid, and level footing for the transport of the heavy loads of concrete. Each dump of a wheelbarrowful of concrete was punctuated by the dropping

HOMEMADE LEVELING METHOD

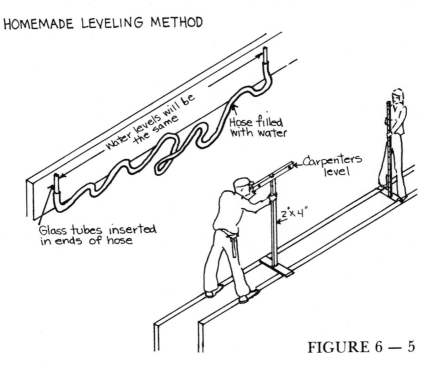

FIGURE 6 — 5

in of several of the filler stones. We took care to see that each of the filler stones was completely surrounded by concrete, and that they did not touch the sides of the enclosing forms. It is a good idea during the course of the pour to check that the filler stones do not protrude above the chalklined level of the top of the footing.

As we progressed with the pour of the footings, we placed the drift pins into the hardening concrete. We placed them at eight foot intervals, spacing them between the standing vertical reinforcing rod (which we also placed on eight foot centers). These eighteen inch driftpins should protrude eight or nine inches above the top of the footing. Their purpose is to prevent the foundation wall from shifting laterally.

If you want to conserve steel costs, you can key the footing (see figure 6-6). This key is made by waiting until the concrete has begun to set — only slightly — and then grooving the top of the footing with a length of 2x4.

Although not imperative, it is a good idea to complete the pour of the footing in one day. To accomplish this with our house, it was necessary to begin at daybreak and to then work through well into the night.

If you must break this process, you should leave the unfinished face of the just-completed pour rough, and it should be banked

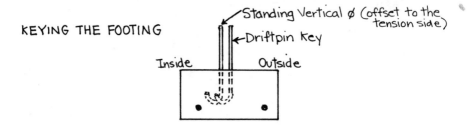

KEYING THE FOOTING

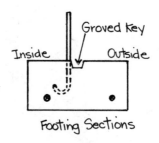

Footing Sections

FIGURE 6 — 6

downward and outward — towards the outside of the wall. Upon resuming the pour, you should begin by covering the old concrete at the joint with a well-mixed batch of cement paste (a straight mix of portland cement and water).

This method for making vertical masonry joints is one that should be followed throughout the construction process — both of the footings, and of the walls. Having these vertical joints appear one directly above another is inviting problems, and you should make it a practice to stagger them at a minimal ten foot distance.

Having your footing done is a landmark in the building procedure. The time it takes for the concrete of the footing to harden (four or five days) should be adequate time for some well deserved self congratulation.

At the end of this hardening time one should get on with the business of placing the first layer of forms for the wall. To get ready for this next step, we removed all the footing forms, and swept the footing off. The sweeping was integral because it is important to obtain as good a bond as possible between the wall and the footing.

To lay out the wall on the footing, we ran strings between the appropriate nails on the batterboards, and then transferred these lines down to the surface of the footings with a plumb bob. We found that gusty winds affected our strings that were strung over long distances, and we therefore deliberately sought windless moments for this operation. Once the proper position of the wall was marked on the footing, we clearly established it by snapping a chalkline. It was then a simple case of making the wall forms toe the line, and periodically checking to make sure that the forms were plumb.

Unlike the subsequent forming layers, the first level requires bracing along its lower length. This bracing was, for us, done with scrap lumber wedged to the sides of the excavation (see figure 6-7). Lower edge bracing is unnecessary on the layers above the first, because their lower edges are secured with bolts — attaching them to forms below.

Another difference between the first and subsequent levels is the difficulty of extracting the first level of forms after the second has been poured, and has set up. To make this removal easier, we inserted shims (scrap wood shingles) under each of the first level forms. When it came time to remove the forms from this level, we pulled the shims out, and this usually gave us just enough "play" to wiggle the forms free.

Once the first level of forms was in place and braced, our next step was to wire them together. The wiring routines we followed are described at the end of the previous chapter. With the wiring done, we inserted the corner reinforcement. This constituted the last preparation work on the forms unless we planned specialized openings that would occur in this level of forms.

FIGURE 6 — 7

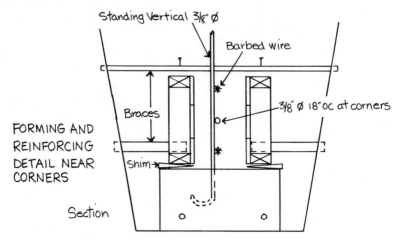

Standing Vertical 3/8" ∅

Barbed wire

Braces

3/8" ∅ 18" oc at corners

FORMING AND
REINFORCING
DETAIL NEAR
CORNERS

Shim

Section

If you are following the progress of this account, you should double check to see that you are providing for sewer outlets, basement drains, waterlines or any incoming electrical systems. We created these openings by nailing in appropriately-lengthed tin cans or scrap water pipe.

Wall levels that are to be below ground level do not require facing stones, but like the footing they can be filled with larger stones to save on concrete. We made it a practice to have these stones ready (cleaned), and stacked close at hand. If, however, you do not have to provide for frost heaving, you will probably want to begin your stonework on the first level of forms that rise from the footing. As was the case with the filler stone, you will want to clean these facing stones, and have them placed nearby. The process of putting the facing stone in place is examined in detail in the following chapter.

Providing for wheelbarrowing was next on our agenda. As we did the footing, we inverted standard forms atop the wall forms as a runway, and we then ramped to this runway from the adjoining excavation. Because the wall forms were narrower than the foundation forms, we found it necessary to construct a slant-sided hopper that would direct the dumped concrete into the area between the forms (see figure 6-8). The hopper saved us the work of a lot of shoveling, and sped up the process considerably.

Dumping the concrete between the forms sometimes tended to create unwanted air pockets, and after one or two mistakes like this we made it a practice to tamp each load with a length of reinforcing rod.

Some notes concerning the quality of the concrete should be inserted here. No matter how carefully you may form or otherwise provide for the pour, the final, most important factor you will confront in putting up the shell of your stone house is the quality of the concrete. The mix proportions involving cement, sand, gravel and water are important factors in the quality of the concrete, but more important is how carefully the ingredients are mixed.

We used a 1:3:4 mix (cement to sand to gravel) with the aggregate coming to the site in an already mixed state — natural bank run. Sue did all the mixing of concrete for our stone structures, and her monitoring of the quality of our concrete is responsible for the proven soundness of our walls.

To describe the wetness of concrete, we could lapse into technical "slump" jargon, but the reality of the building site makes such discussion hopelessly academic. Be guided by this adage: *When in doubt, make your mix dry!* There is a tendency to keep adding more water to get a more "plastic" mix. Resist that tendency, and you will not be sorry later.

We found that the best mixes were the ones where the dry ingredients were mixed first. This held true whether the mix was made with a motor-driven mixer or with a simple trough. Adding the water afterward made for a longer mixing time, but the results were more satisfactory. The wetness of our sand and gravel piles varied widely, depending upon the weather, and therefore the amounts of water required for batches fluctuated correspondingly. This fluctuation saw us using anywhere from three-and-one-half to five gallons of water for each bag of cement used in the mix.

Lest we scare the reader off, it should be said that concrete is a relatively undemanding medium, considering its final permanency. A

Hopper for pouring foundation wall

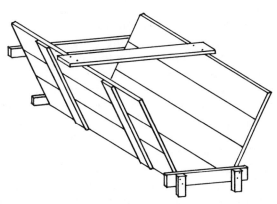

FIGURE 6 — 8

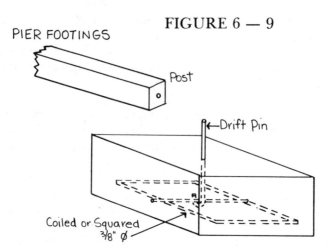

FIGURE 6 — 9

PIER FOOTINGS

Post

Drift Pin

Coiled or Squared
3/8" ∅

friend who lives down the road from us, and who has had considerable experience with concrete once remarked, "Cement (Vermonters do not distinguish between mortar, concrete or portland cement) is awful funny stuff. It's like a broody hen — it doesn't take kindly to being moved once its begun to set."

Careful preparation for the pour, and anticipation of the inevitable delays between the mixer and the form end of the operation will prevent a lot of waste, and save on worn nerves. If delays cannot be avoided in placing the concrete, you may have to add more water to the initial mix, but do not try to add water to a batch that has stood for any length of time.

Once off the footing, we found that our routines for forming and pouring became so second nature that we were able to form and pour a section of wall eighteen inches high by thirty feet long each day. At this rate, we made very satisfying progress. It seemed no time at all before we were up to the third level of our forms.

About the time we reached this height on our walls, we decided that we would have to pour the pier footings and the basement interior walls before we had boxed ourselves out. The pier footings were relatively easy to pour, and required only careful mensuration to determine where they were to be placed. We formed them at the precise spot where they would rest to support the posts and the girders.

As the pier footings are not exposed to the ground freezing possibilities of the wall footings, they do not need to be excavated. We formed and poured them directly on the subsoil that was uncovered when the site was stripped. Reinforcement of these footings was done with 3/8-inch reinforcing rod arranged and wired in the fashion shown in figure 6-9.

Because we had elected to have a partial basement, we had to form and pour interior basement walls. Like the exterior walls, they were also footed, but because they were virtually non-support structures, we formed and poured them to a mere six inch width. They were, however, widened and reinforced where they supported more than their own weight. (Two points on these walls were incorporated to support floor girders.) We found that the time it required to form and pour these interior walls was a strong argument for full basements — basements that extend under the entire house.

Full basements have other positive attributes that outweigh the increased costs of excavation and concrete for pads. Primary among these attributes is the additional floor space obtained. Floored with a concrete pad, they become extremely flexible in usage. Non-support partitions are easily erected to divide the area into useable units, and these units can, with a minimum of effort, be shifted and moved about to accommodate your changing needs.

Whether you choose a full or a partial basement, we recommend that you floor the entire area with a concrete pad. You will have provided for this pad when excavating, and these provisions will have included one or more floor drains.

While you may not have an obvious or immediate need for these drains, their expense and trouble of installation will eventually justify themselves. We use our basement to house our systems and machinery, and we have set aside space there for wood storage. Your needs may differ from ours, but the basic procedures for forming and pouring a pad are the same. Figure 6-10 shows how we constructed ours.

BASEMENT PAD
Section

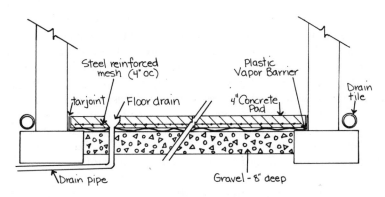

FIGURE 6 — 10

We first hauled and spread eight inches of gravel in the space provided by excavation (to assure good drainage), and then covered this with a heavy plastic vapor barrier. Over this we rolled out steel reinforced mesh (four inches on center). Rolls of this mesh are seldom wider than five feet, and the narrow width necessitates overlapping and wiring. This reinforcement is often omitted in pads where there is little likelihood of the pads being subjected to heavy loads. Concrete pads will be very cold underfoot in *any* event, but you can make them somewhat warmer by insulating their juncture with the exterior walls. We used fiberglass sill sealer for this purpose. Then after the pour was made and had set up, we finished the joint off with tar.

The above preparation found us ready to pour the pad. The surrounding walls served as forms, and on them we snapped a chalkline to indicate the height of the finished pad where it would contact the walls. To guide us in the desired depth of concrete in the middle of the pour, we pounded in several grade stakes. These were removed as we made the pour, and they also served to remind us to lift the reinforcing mesh so that the wire had at least one or two inches of concrete beneath it.

Most importantly, however, the grade stakes marked the incline of the finished floor that slanted towards the imbedded drain. If you follow this same procedure, you should remember that when working the concrete with screed and float, you will have a great deal of water working to the surface. Unless you plug your drain(s) before you begin, you can be assured that they will be plugged when you finish.

Pouring a concrete pad within a confined area like a basement is a gunky job. On our first job we both ruined our boots, and have since approached similar tasks clad in washable rubber boots. Clothing is replaceable, hands are not. Working steadily with concrete without sturdy rubber gloves will result in sore hands (at the very least), or even "cement burns", split fingertips and potential infections. Our short building season did not allow us the "luxury" of time taken out to heal, and we wore these gloves whenever we approached concrete.

We also found that these gloves were excellent for handling stone, because they were heavy enough to ward off the minor abrasions and pinching that is inevitable when putting stones between forms. They also seem to help in handling wet, slippery stones by providing a firmer grip. The only drawback we have encountered with these rubber gloves is that they do not "breathe". Periodically we had to take them off to rest our hands, and to air their insides. They acquire an objectionable odor after a while, but you yourself will not be too fragrant after a few hours laboring over the forms.

Before you get too high with the exterior walls you will want to backfill the excavation. If you live in a perenially wet area, or if your soils are poorly drained, you will want to put down drain (weep) tile. Putting this tile around your house will require you to top it with a two-foot or so layer of porous gravel. Having no such drainage problems is a blessing, as you can simply replace the excavated dirt around the footings and foundation wall.

When finished with the backfill, there is an appearance to the site that the house has finally emerged from the ground — as indeed it has. But a more practical plus is the fact that you will not have to swear your way around the impediments of mountainous piles of dirt. We took advantage of this landmark time to clean up around the site, and to prepare for the big push to get the remainder of the exterior walls done to the plate.

Foundations:
"Whoeso diggeth a pit shall fall therein."

Proverbs, XXVI.27

* * * * * *

"When the Roman stoic experienced catastrophes, he took them with courage of resignation. But the typical American, after he has lost the foundations of his existance, works for new foundations."

Paul Tillich

Exterior Walls

Chapter 7

In stone houses having a crawl space, the exterior facing stones appear in the wall well below the first floor level — the exception to this is where a house is built into a hillside. If the house site is on a flat, level piece of terrain, the stone line is easy to visualize, but our site is, for example, a rolling one requiring slanting grades and multiple terracing, and its stone line is a complicated one. It was particularly difficult to establish this line when building the exterior wall, because the forming process involved several different levels of forms, each rising at a different rate from its neighbor.

Early on we found that it was easy to miss the spot where the stonework should begin. With this discovery, we then made it a standard practice to flag suspected levels with colored tape. If we then determined that the flagged level would be one where the earth would lie against the wall, we guessed at the grade level, and then marked it as a guideline with a snapped chalkline.

The basic rule of stonelaying seems so simple — indeed, obvious — that it hardly seems worth repeating, but the history of stone shelters is rubbled with testimony to the ignorance of our disdain for this rule. The rule is this: ONE OVER TWO, TWO OVER ONE.

This means that the stonelayer places one stone atop two in such a way that it bridges them, and then, later, places two stones over the first one: see figure 7-1. This same figure shows two patterns for the laying of stone. You should note that both coursed stonework and random stonework observe the basic rule.

The two obvious benefits that accrue from this rule are stability and strength. The fact that slipforming provides a concrete backing to stonework does not negate the need for observing the rule — to the contrary. Since the fieldstone in our house occupies over one half the volumetric space of our walls, it would be folly to ignore it.

To avoid shoddy stonelaying, it is a good policy to have an adequate supply of stones from which to choose. What quantity you will

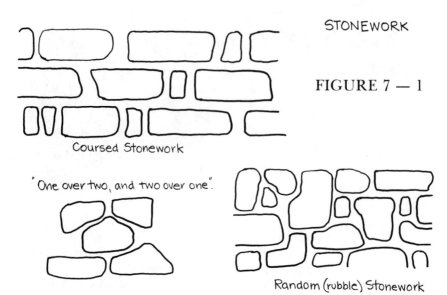

STONEWORK

FIGURE 7 — 1

Coursed Stonework

"One over two, and two over one".

Random (rubble) Stonework

require depends upon the general size of your stones, but the experience of one or two pours will usually give you a fair idea of what to expect. We soon found that we required different stones for different zones, and while we were getting enough stones together for a specific pour, we made it a point to have the kind(s) of stones available that the pour called for. Generally, the stones fell into four categories. (Fireplace stones, keystones, lintel stones and centerpoint stones are discussed elsewhere.)

Wall stones:	These were the most common stones used in our walls. They came in all sizes and shapes, but they bore two characteristics in common: 1) they all had at least one flat surface that could face the wall, and 2) they all had a second dimension that was at least one or two inches less than the width of our wall.
Stretcher stones:	These were stones that we used to bridge two or more wall stones (in pattern). They were usually placed horizontally in the wall, and were, in general, oblong in shape. Occasionally these stones served as quoins.
Quoin stones:	This is the formal name for cornerstones. The requirements for this stone were that they have at least two flat surfaces that met at something approximating a 90 degree angle. Because there is somewhat more form space at the corners, these stones could be somewhat larger than wall or stretcher stones.
Reveal stones:	These are specialized quoin stones used to face the interior (and exterior) of window and door reveals.

We chose to build our house using the random stonework patterns. For us, working with the stone that came from our own fields was a particularly satisfying experience. This choice was a pragmatic one that arose from the fact that we had an abundance of native fieldstone.

It should be noted that fieldstone, with its irregular shapes, does not lend itself to coursing. While the irregularity of stone shapes sometimes posed a challenge during the building, this same irregularity was also responsible for a desirable spontaneity, both in the building process, and in the finished product.

Looking back, we can see that using fieldstone from our fields was a fortuitous choice. Comparing other stone houses that used "alien" stone with our own, we are struck with the organic wholeness of ours as it sits in the midst of our fields. Other stone houses using cut stones or those far removed in origins from their resting place, appear to us to have a transient, out-of-place look.

Stonelaying between slipforms presents some problems that are not encountered in the more traditional handlaying methods. The most vexing of these is the fact that you must lay your stones up backwards — that is, their faces are against the forms, and you cannot see the actual face of the wall as you lay it up. We found that this drawback could be overcome with experience. This means that it took one or two levels before we got the "feel" of the wall. After we had poured and removed the forms of a level, we got a sense of continuity to the wall, and the mechanics of placing the stones so that they made the pattern we wanted seemed to come quite naturally.

One of the stonelaying routines we established was to place a whole row of stones between the forms immediately before we were ready to pour. In cases where it was obvious that concrete would not flow around the base of a stone, we pulled it before shoveling the concrete in, and then replaced it.

This "dry" run allowed us to assess the stones that were already imbedded in concrete to see where a stretcher should be placed, or to assess the proper overlap for quoin stones. More often than not, we found that this initial row of stones could be left in place during the pour — providing that the faces of the stones were wedged against the outer form, and that we carefully tamped the wet concrete to see that it flowed under the base of the stones.

For wedging purposes, we first tried scrap pieces of wood, but found that we occasionally forgot to remove them. (This is the surest way to leave an unwanted hole in your wall.) Thereafter we used stone chippings or shards of larger stones for this purpose (see figure 7-2).

One mechanical footnote should be added here: when placing the

FIGURE 7 — 2

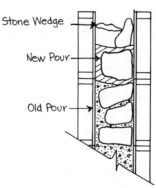

final stones of a pour, we always checked to see that no stone protruded above the face of the outside of the form. Once a protruding stone sets in concrete, we found it impossible to do anything but chip it off with hammer and chisel before we could set the next level of forms in place.

The most common error slipform stonelayers make is that of putting their stones too close together. This detracts from the wall both esthetically and structurally. On the outer face, a two-inch pointed joint is a good proportional spacing to aim at, and it provides an adequate "glue" between the stones.

To this end, we found that it helped to pile the stiff concrete against the inside form before putting the stone into place against the face. This allowed us (Karl) to hold the stone in place with one hand, and to sweep the supporting concrete into place with the other. The person handling the stonelaying should make it a point to raise the wall at an even rate so that he can maintain both horizontal and vertical continuity in the stones being laid.

As a general rule, the largest stones in a house wall are found nearer the bottom of the house — a sense of proportion that arose from dry laying practices, and which the human eye now seems to find pleasing. It is also a fortuitous happenstance, as it makes lifting the stones up for the higher layers a much easier job.

You will notice that to this point we have not discussed the handling of larger stones. The reason for this is that there is no need for using stones weighing over 40 or 50 pounds. It was a rare occasion when Sue could not heft a stone that we used in any of our structures, and unless you have no other alternatives (*i.e.* you have no stones other than large ones), we would suggest that you restrict your weightlifting to the gym.

But even with the relatively light weights of average wall stones, you should exercise common sense — lift with your legs, not with your back. We found that the hardest physical part of stone moving occurred when the stones were being finally placed between the forms. It is at this point where one's body is at a disadvantage, and that the potential for muscle strains is highest.

Aside from the normal, work-induced muscle aches, we incurred no back problems, and have survived several house buildings with no displaced discs or other physical disasters.

Once the level of the wall begins to show stones on its face, and you have backfilled, you will find that all of the work will go more smoothly. We took time from the walls at this point to put our floor together. For this we addressed our attentions to the interior of the outer walls.

Here we made preparations for joining the underpinnings and the floor with the exterior walls. We joined our underpinnings (girders) to the wall with four-inch deep pockets, which were set into the wall with simple forms made from scrap lumber (see figure 7-3).

Our floor joists were set on 2 x 4 sills that occupied a shelf created when we narrowed our 14-inch foundation walls to 10-inch exterior walls. We fastened these sills to the shelf with bolts set head-end in the masonry, and then toenailed the joists to them. Since there was no header to space the floor joists (and to keep them from twisting), we inset blocks of lumber the same size as the joists between them (figure 7-3). This sill gave us a four inch purchase (a minimum bearing surface) for our 2 x 10 floor joists, and we took some pains to see that the sill was level and straight.

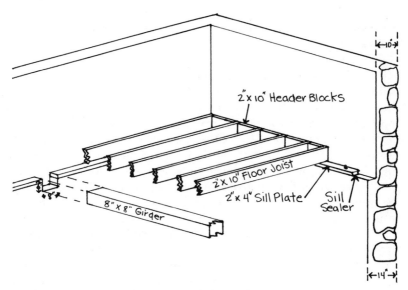

FIGURE 7 — 3

The inset sill can cause some difficulty in forming if you do not give the matter a little thought. In our case, we knew that we wanted a basement having a minimum overhead clearance of seven feet, six inches, and on scratch paper we calculated that five levels of 18½ inch forms would, if we filled and smoothed the concrete to the top of the fifth level, give us more height than we wanted or needed.

We compromised by smoothing the sill inset two inches below the top of the interior form, which, when silled and joisted, gave us a basement overhead that was seven feet, eight inches above our four-inch pad.

After the smoothed inset was implanted with bolts and allowed to set, we cut and rolled out fiberglass sill sealer, put the 2 x 4 wooden sill in place and ran the nuts down on the bolts. The next level of the interior form was then secured to this sill, and the pours continued with the inner and outer forms at the same level.

The day finally came when we began the assembly of our floor. We first cut our posts to set on the pier footings, and got them in place preparatory to cutting the girders. For the posts and girders we used some rock-hard eight-by-eight beams taken from an old barn we had salvaged on our farm. We had barely enough timbers to do the job, and were being exceedingly careful in measuring and cutting.

George, the name we had applied to a resident pet chipmunk, would come to watch the operation daily, but only after the morning's sun had warmed the site. Cadging food was George's only tie to the human newcomers, and he would alternate between stuffing his cheeks with acorns from an adjacent white oak and bread crumbs set out on the inset sill for his benefit.

The morning that we cut the main girder, George perched on the sill watching us as we sweated and strained to get the girder into place. It slipped into the pocket (we allowed ½" space for "slop" in the forming) in the wall beautifully, but when we lifted it to the top of the appropriate post, it was about two inches longer than our plans called for. Since we had already poured the pier footings, we suffered some bad moments as we considered the alternatives. Had we mis-measured in placing the piers? A check proved the piers were where they belonged.

We had just about given up, and resigned ourselves to cutting the beam when we discovered the cause of our problem. George had been using the girder pocket as a repository for his acorns, and as we cleaned out his cache, we also discovered a plumb bob that we had "mislaid" two weeks earlier. I don't think that we ever poured George into the wall, but at the time we gave it some serious consideration.

After we had the sub-floor in place, we used it to brace our interior forms and to run the wheelbarrow on. After slogging through mud and over soft, piled backfill, the smooth, stable underfooting of the sub-floor was a pleasurable change.

From this point on we did all of our stonelaying and pouring from the interior of the house. The sub-floor also provided us a reliable base on which to foot the staging we required for the final two levels of forms.

Our house, with its low seven-foot wall plates and wooden gables, required minimal staging. We acquired two large wooden reels from the local telephone company and joined them with two-inch planks. This constituted all the staging we found necessary, and they were quite portable.

We did not stone the gables on our house, but in subsequent buildings where we did, we discovered several common problems (and solutions) that would certainly apply to any gabling.

We found, for example, the problem of staging to be a serious one. Our barn gables were particularly difficult as they rose fourteen feet above the barn pad. Staging for these heights required the erection of 2 x 4 structures that were nowhere as stable as the telephone reels, and took some time to construct.

We wheelbarrowed our concrete to these stages by ramps where we dumped it into a "holding" trough. When we had the help of the

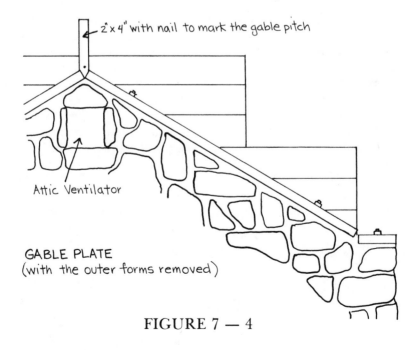

FIGURE 7 — 4

occasional visitor — (visitors were put to work whenever possible) — we had one person on the cement mixer, one person doing the wheelbarrowing, and one person on the forms doing the final shoveling of the concrete (from the holding trough to the forms) and laying the stone.

We joined our gable ends to our rafters with a plate that was bolted to the wall in the same fashion as the rest of the wall (see figure 7-4). To determine the roof pitch as it would be poured in the gable forms, we erected a long 2 x 4 at the point where the roof crown (ridge pole) would meet the facets of the roof over the gable.

Then, driving a nail into the 2 x 4 at this meeting point, we snapped a chalkline betwen the nail and the top of the exterior wall. In progressing up the gable end with each pour, we snapped this line for each level, and were careful in following it that we did not let any stones protrude beyond the line. More often than not, we stopped the stonework two or more inches below the line, and filled the remaining distance with concrete. The latter had to be of a stiff nature in order to prevent slumping bulges, and once put into position we made it a practice to work it as little as possible.

Near the peak of each gable we built in attic ventilators. In the wood gables this was a carpentry job, but in stone gables the problem, like that of crawl space ventilators, was to pre-form the openings in the wall.

Ventilators are probably the easiest of these openings in the wall to form — unless you fancy arches or other difficult lintel work. We chose wooden rectangles that would accommodate louvered and screened aluminum ventilators. These ventilators are standard stock at most building supply stores. The rectangles were made of one by nine boards, which we cut and butt-nailed together.

Once inserted between the forms, they were intended to stay in the wall, and so we treated them with a preservative and "porcupined" them with nails taken from our bent, salvage nail supply. Figure 7-5 shows

VENTILATOR FRAMING

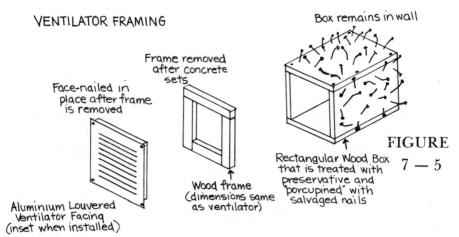

Box remains in wall

Frame removed after concrete sets

Face-nailed in place after frame is removed

FIGURE 7 — 5

Rectangular Wood Box that is treated with preservative and "porcupined" with salvaged nails

Wood frame (dimensions same as ventilator)

Aluminium Louvered Ventilator Facing (inset when installed)

this process in detail. This same figure shows that we framed the face of this rectangle with one inch scrap lumber.

You will notice that this frame is *not* "Porcupined". This is because we wanted to remove it after the forms had been taken away, and the concrete had set up. The purpose of this frame is, when removed, to provide a one-inch deep negative impression in the concrete. Then, when put into place, the ventilator is recessed enough to shed the effects of weathering, and it overlaps the interior rectangle of wood so that there are no "through" joints.

No through joints is the basic rule for putting openings in any wall, but it is a particularly applicable rule with stone houses. Once set, concrete is unforgiving. It is essential, therefore, that extra precautions be taken to eliminate direct joints — those from the outside to the inside of a stone wall.

The most likely place where these mistakes can occur is around the framing for wall openings, and you must plan in interruptions or baffles to take care of the space that occurs when the concrete dries and shrinks. As it shrinks, the concrete pulls away from adjacent wood framing, and, unless you have baffled the space, it will allow weather to penetrate.

There are several ways to make these baffles. Anyone who thinks about it can probably come up with as many ways to do the job as we did. Some of the ways we tried in various buildings are shown in figure 7-6. The method you employ will probably depend, in some measure, on what kind of framing materials you are working with. We chose large 8 x 8 timbers for framing around all our doors and windows. These timbers were the same that we used for our girders (old barn beams), and because they were hand-hewn we were forced to make some adjustments to accommodate their rough faces.

Simply stated, what we did with our 8 x 8 beams was to pre-assemble them (one large window had to be put together in place), and then lift them into their resting spot between the forms. We needed extra help to do this as most of the frames weighed well over 200 pounds by the time we had braced them and they were ready for the forms. When set in place, they were flush with the interior form, and because the wall was ten inches wide, there was a two inch gap between the 8 x 8 frame and the outside form.

We filled the vertical gaps with 2 x 4 stock (we did not cut the 2 x 4's to length because we reused them later). The gap around the bottom side of the frame was left open so that we could be assured of being able to work the wet concrete completely under the frame. When the concrete completely filled the area under the frame, we built up a concrete sill (apron), which we slanted downward with a trowel to assure good drainage.

FIGURE 7 — 6

FRAMING OPENINGS

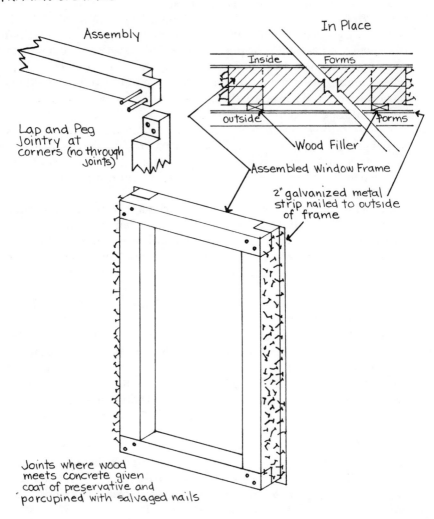

Assembly

Lap and Peg
Jointry at
corners (no through
joints)

In Place

Inside   Forms

outside   Forms

Wood Filler

Assembled Window Frame

2" galvanized metal
strip nailed to outside
of frame

Joints where wood
meets concrete given
coat of preservative and
'porcupined' with salvaged nails

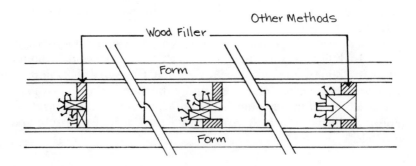

Other Methods

Wood Filler

Form

Form

To check the possibility of air leakage due to shrinkage of the concrete, we formed the frame in such a way that concrete covered about two inches of the front of the frame on the outside (see figure 7-6).

Then to make this joint completely secure, we nailed a two inch strip of galvanized metal, cut from a sheet of roofing, to the outside of the frame. The only difficulty that we encountered with this system was that it limited the choice of reveal stones to a narrower dimension than we preferred.

Because nearly all of the window and door frames used in our house ended at the top of our walls, we did not have to contend with lintels that would have to support masonry too. Figure 7-7 illustrates how the wall plate crossed our window and door frames. This jointry has proven satisfactory, but we found the need, in later construction, to install lintels that would bear masonry above them. Our efforts to date have been restricted to rudimentary wooden beam lintels, reinforced concrete lintels and single lintel stones.

Historically the stone lintel is probably the oldest method known to man for topping a door or window area. It is the simplest solution to the problem, as it involves but a single stone that spans the erected sides of an opening. The simplicity ends here, for this stone must support weight. How much weight it can support depends upon the thickness of the lintel, the distance it must span and the nature (type) of stone being used.

As was pointed out earlier, our stones were of the metamorphic variety having a high compressive strength and a relatively low tensile strength. This tensile strength would, of course, increase as the thickness of the stone increased, and it was owing to this fact, in part, that we limited our use of lintel stones to spans of three feet or less. Spans greater than this called for stones that we felt were out of keeping with the scale of our structures.

Individual stones within a type differ markedly, and it is therefore difficult to give the reader hard and fast rules for estimating the required thickness for a specific kind of lintel stone for a given span. Hidden

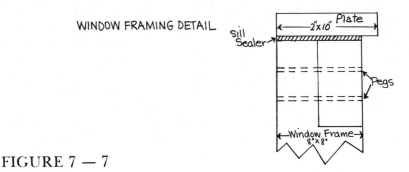

WINDOW FRAMING DETAIL

FIGURE 7 — 7

CALCULATING STONE LINTELS

FIGURE 7 — 8

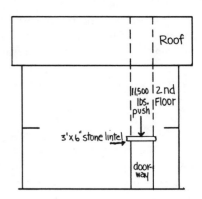

fracture lines or other unsuspected imperfections will alter the most precise equations you may work out, and the end result could be ruinous.

Having pointed out the worst that can happen, we hasten to add that many of the world's oldest structures have stone lintels. We would also add that in our view, stone lintels or arches lend immeasureably to the sense of wholeness of a stone house.

To calculate the thickness of a lintel versus its required span, you must first know the weight it is expected to bear. For figuring purposes, we took that area directly above the lintel (including the proportional share of the roof and 2nd floor), and converted this area to pounds — we used the weights indicated on figure 4-6, Chapter 4. The 30-inch doorway shown in figure 7-8 requires a three-foot lintel, and this lintel is expected to support, at a generous estimate, 12,000 pounds.

As an example, say we use one of our metamorphosed granite stones as the lintel, and assume it to be six inches thick and one foot wide. This six-inch by twelve-inch by three-feet long lintel will support roughly 13,000 pounds (uniformly distributed weight). A foolish builder might consider this an adequate margin.

We say "foolish" because by merely increasing the thickness of the lintel by two inches its potential load capacity is nearly *quadrupled*. To err is human — to estimate on the side of safety is divine. This maxim holds true for house building and poker . . . . love and monopoly are something else.

Table 7-9 contains the co-efficients of *minimum* transverse strength used to calculate the breaking load when concentrated at a point midway on the span. A load uniformly distributed (as is the case with most houses) over the span of the lintel doubles the maximum potential load weight. To use this table, apply this formula to a proposed lintel stone.

$$\frac{\text{width (in inches) x thickness}^2 \text{ (in inches)}}{\text{length (in feet)}} = \text{X TVS co-ef-} = \begin{array}{c}\text{maximum}\\ \text{concentrated}\\ \text{load weight}\end{array}$$

To the eight inch thick lintel described above, the formula would be applied as follows:

$$\frac{12 \times 64}{3} \text{ X } 100 = 25{,}600 \text{ pounds (maximum concentrated load weight)}$$

$$\frac{\times 2}{51{,}200} \text{ pounds (maximum uniformly distributed}$$
$$\text{load weight)}$$

A useable lintel stone is hard to find, particularly since it must have a dimension that will fit between the slipforms. Because we sought to use native, unquarried fieldstone, this search has become even more difficult for us, and we find that we are ranging farther and farther afield for these prized stones.

Our tolerant neighbors have become inured to our prying and poking about old stone walls, and some have even joined us in our "madness." Last summer a neighbor gave us a lead on a possible lintel stone. It was one that he had unearthed in a remote stone wall while out walking a property boundary. He offered it to us, and we dropped everything to go have a "look-see".

It was an outright jewel of a stone, and we were determined to get it back home. We bartered with another neighbor for the use of his horse, as the section of wall where the stone lay was long overgrown, and there was no way to get a motorized vehicle in to the spot. After much travail we got it out to where we could load it onto our pickup, and with some smug satisfaction drove the remainder of the way home. In unloading the stone we dropped it, and it split dead through the center.

TABLE 7 — 9

Transverse Strength of Stone

Kind of Stone	Co-efficient of maximum transverse strength		
	Maximum	Minimum	Average
Metamorphic granites	150	50	100
Limestones	140	8	83
Marble	160	8	120
Sandstone	130	32	70
Slate	500	100	300
Blue-stone flagging	251	20	150

We are still paying off on the barter, but we shrug it off, assuming that it is all part of the dues paid for being card-carrying stone nuts.

Other types of lintels we have used were a wooden beam (8" x 12"), and a reinforced concrete beam. The former seemed to be of doubtful durability to us, and we took the additional precaution of inserting reinforcement (an old John Deere tractor axle) in the masonry directly above the beam. There was little masonry above the opening, and the beam's load bearing potential seemed more than adequate.

The reinforced concrete beam lintel was cast in place over an extended strip window. This method boasted two very satisfying results: 1) the presence of the reinforcing rod was (and is) a comforting reassurance concerning the lintel's tensile strength; 2) the reinforcing rod did not interfere unduly with placing wall stones on the face of the lintel. It is this same technique that promises to make the construction of arches safer, surer and much more easily done than with the traditional unreinforced stonework.

ARCHES

The following discussion of stone arches is included for the reader's (and the writer's) interest and edification. We have helped with the construction of one small segmental arch, but, at this writing, have never undertaken an arch of our own. We are, however, making plans for our first effort.

Arches have been around for over 2,500 years, and yet no one, to this date, knows precisely how they work. All in all, the history of arches is an incredible and fascinating study in theory — and faith. If you doubt the latter, just wait until you knock the scaffolding out from under your own first effort.

Theory has it that the vertical forces of weight above the arch are "bent" horizontally, and thus transmitted to the foundation work via wedge-shaped stones (archstones, ringstones) that are arranged in a curve.

The attempt to account for (and thereby to control) all the forces at work in stone archwork has occupied mathematicians and engineers for thousands of years. These specialists have spawned as many theories as Bayer has aspirin, and every hundred years or so their theories are completely revised as someone builds a bigger, longer, more impossible arch.

Where does all this leave the stone house builder who modestly aspires to an arched window or entryway? Considering the history of successful arches (the failures unrecorded?), we would assume the odds are pretty good. One old treatise on masonry construction we came

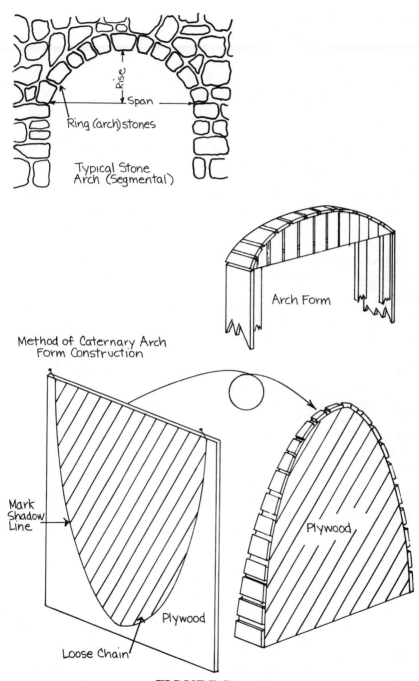

Rise

Span

Ring (arch) stones

Typical Stone
Arch (Segmental)

Arch Form

Method of Caternary Arch
Form Construction

Mark
Shadow
Line

Plywood

Plywood

Loose Chain

FIGURE 7 — 10

across in our research devoted 150 pages to complex theory and formulae of arch construction, and then concluded:

> ". . . . the stability of a masonry arch does not admit of exact mathematical solution, but is to some extent an indeterminate problem . . . Considered practically, the designing of a masonry arch is greatly simplified by the many examples furnished by existing structures which afford incontrovertible evidence of their stability by safely fulfilling their intended duties . . . In designing arches, theory should be interpreted by experiences . . ."

Experience would suggest that constructing a reinforced arch within formwork would be only slightly more difficult in practice than the methods already cited for other openings in the exterior wall. Instead of a door frame, you would substitute an arch scaffolding (see figure 7-10). The stones would be put between the forms differently in that the ones lining the arch (faced against the scaffolding) would eventually be exposed.

Stone selection for the ring stones would entail the same rationale as that invoked for the traditional handlaid wall. That is, the stones would have to be wedgeshaped, and of such configuration that they would bear on one another with near-matching surfaces. Where traditional arch work relied strictly upon stone selection and mortar for its structural strength, a formed arch would provide a solid concrete backing into which the builder can imbed reinforcing rods.

Functionally the arch is superior to the lintel stone in that it will accept greater loads. While it historically boasts other functional advantages, these advantages do not particularly benefit the ordinary stone house builder.

Because we have no heavy wall weights to support, and harbor no ambitions to create airy vaults or classic domes, we regard our first forays into archwork as esthetic adventures. To our way of looking at it, the soft, rounded lines of our stones go well with the romantic curves of the arch.

POINTING

For many builders, the esthetics of a stone house outweigh its functionality, and for them the final, finishing work on the face of the exterior wall is all-important. This finishing work is called pointing, and it entails filling in the cracks, crevices and joints between the stones with a rich mortar (three parts sand to one cement, or even two to one).

In our experience, this exercise is not a functional necessity. The dire warnings that cold weather would, by expansion, crack stones from our unpointed wall have just not happened, despite the fact that our climate sometimes sees temperatures drop to -30°. But if you feel the need to point your walls, you should begin before the concrete sets up. The best time to start is immediately after removing the outer form. This means that you must interrupt your upward progress at each level.

Whether you decide to point or not, if you have followed the progress of this chapter, you will have completed the outer boundaries of your home. For us, this was a particularly important time. Our building season is a short one, and because we began late in the building season, we pushed to get our exterior walls up before freezing temperatures set in.

Concrete can be mixed, and it will set in freezing temperatures (with the aid of chemical additives like calcium chloride), but it is not recommended practice. With our walls erected, we set about framing the roof in order that we might make our shelter a shelter.

"So built we the wall . . . for the people had a mind to work."

Nehemiah IV.6

𝓕raming & Skins
Chapter 8

E xcepting the basement pad, the horizontal surfaces of our house were made of wood. While nowhere as durable, nor in our view as esthetically pleasing as stone, wood is a flexible, forgiving kind of medium. It can be tacked together to cover innate weaknesses, untacked to correct human error, and otherwise altered and amended to suit the needs of the builder.

When used for joists and rafters, it must be at least two inches thick. The necessary width varies depending upon: 1) the spacing between joists or rafters, 2) the span (the unsupported length), 3) the loads the span is expected to bear and 4) the kind and grade of wood being used.

Table 8-1 indicates the maximum spans over a given spacing for floor joists, ceiling joists and roof rafters. These spans are dependent upon top grade lumber being used. If you must use lumber with structural defects, and most of us do, it is wise to lessen the OC (on center) spacing, and/or decrease the length of the span.

Floor and Ceiling Joists

Joists, by definition, reach from sill plate to sill plate, or from sill plate to girder or beam. They can be either floor or ceiling joists, the differences being that floor joists carry their loads on their upper surfaces, and ceiling joists carry theirs on the lower. Both are assigned live and dead loads for purposes of structural calculations. The latter includes the weight of joists and floors, and the former that of materials and people (see figure 4-6, Chapter 4).

As a general rule, floor joists are made from 2x8's, and ceiling joists 2x6's, but the builder should consult the span and spacing table included here for specific situations. Nearly every piece of lumber has a convex and concave side when viewed from its narrow dimension. The

TABLE 8 — 1

FRAMING SPAN AND SPACING

Lumber	Spacing OC center to center	Floor Joists Maximum Clear Span (w/o ceiling plaster)			Ceiling Joists Maximum Clear Span	Roof Rafters Maximum Length	
		40 lbs live & dead ld	50 lbs live & dead ld	60 lbs live & dead ld	30 lbs live & dead lds	20 lbs live & dead ld	40 lbs live & dead ld
2 x 4	12″				9′5″		
2 x 4	16″				8′2″	9′9″	7′4″
2 x 4	24″					8′0″	6′0″
2 x 6	12″	13′2″	12′0″	11′1″	14′4″		
2 x 6	16″	11′6″	10′5″	9′8″	13′0″	14′11″	11′4″
2 x 6	24″	9′6″	8′7″	7′10″		12′4″	9′4″
2 x 8	12″	17′5″	15′10″	14′8″	19′6″		
2 x 8	16″	15′3″	13″10″	12′9″	17′9″	19′8″	15′0″
2 x 8	24″	12′6″	11′4″	10′6″		16′4″	12′4″
2 x 10	12″	21′10″	19′11″	18′5″	24′9″		
2 x 10	16″	19′2″	17′5″	16′1″	22′6″	29′6″	18′10″
2 x 10	24″	15′9″	14′4″	13′3″		24′7″	15′7″
3 x 8	16″	19′1″	17′4″	16′0″			
3 x 8	24″	15′9″	14′4″	13′3″			
3 x 10	16″	23′10″	21′9″	20′2″			
3 x 10	24″	19′10″	18′1″	16′8″			

1. This table assumes the highest quality woods being used (minimum fiber stress of 1200 lb.). Individual kinds of wood (species) will vary in their capacities.
2. These spans permit the attic to be used for minimum storage.
3. Rafter lengths given here are based on slopes greater than 3 in 12, and the length given is the distance from the plate to the ridge. Slopes less than 3 in 12 should be calculated on the same basis as floor joists.

convex side is called the crown, and it is good building practice to install all joists and rafters with their crown sides up.

Old-style stone houses saw the joists inserted into preconstructed pockets in the exterior walls. We used this method for situating our girders, but discarded it as impractical for our joists, owing to their numbers, and the difficulty of forming the pockets. If the reader concludes that he would like to use this method of tying his floor to his wall, he should observe the precaution already cited for girders — that of tapering the wood where it enters the pocket.

A few summers back we visited the ruins of an old California stagecoach stop. It had been built of stone, and the remaining stubs of joists protruded from the old style wall pockets like pointing fingers. The blackened remnants bore mute testimony to what had transpired in this remote foothills waystop, but as we scuffed our way around the ruins we stumbled (literally) onto a mystery. The way the building had fallen made no structural sense. To the east and west the ends of the building remained standing, but the north and south sides where the joists joined the wall had tumbled — to the outside of the building! It would, we reflected, have made more sense if the walls had remained standing, or that they would have been pulled inward by the fire-gutted roof as it collapsed.

The mystery was solved as we looked closer at the joists and the pockets into which they had been inserted. A gaping fracture line stretched from joist pocket to joist pocket, and when we peered into one of these pockets, we saw that the joist ends were *square cut*. When the burning roof collapsed, it had struck the floor joists, and they had operated like enormous levers to pry the wall outward.

In the previous chapter, we briefly described how we joined our floor to our wall. To recapitulate, we set our joists on a 2x4 sill-plate which was, in turn, bolted to a four inch inset in our concrete and stone wall. This inset marked where the width of our wall diminished from 14 inches at its base to 10 inches at the inset. As was described, we then spaced our joists at the sill plate by inserting solid wood bridging (same stock dimensions as the joists) between the joists.

Another variation to this method is to butt-nail a box header to the joists, and then set the whole thing on the sill plate. We used a 1x10 header for this, in order that we could give as much bearing surface to the joists as possible. Where there was less than four inches of joist bearing on the sill plate, we added a ledger to make up the necessary difference (see figure 8-2).

A friend who also used this method to space his joists went on to incorporate the 1x10 header as an interior form for his wall. This necessitated his juggling the inner-outer form levels to assure, in

FIGURE 8 — 2

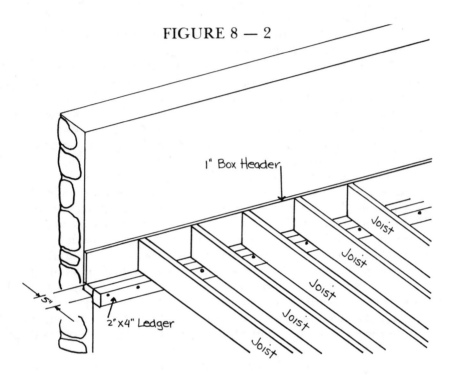

subsequent pours, that the tops of the inside forms were close to the same height as those of the outside.

Our house spans required instances where the other end of the floor joists rested on a girder, and instances where they spanned an entire width from outside wall to outside wall. When they overlapped on a girder, we ordered our joist stock from the sawyer in lengths that required no sawing. For us, a few extra inches of overlap at these junctures was acceptable — even desirable, as they gave Sue considerable practice in hammering them together with 20d nails.

The reader should be warned that hammering contests are an inevitable outgrowth of an extended framing session. Sober, mature adults are reduced to gibbering teenagers as they vie to see who can sink a 20d nail in the least number of strokes — with the least number of "hickies" (mis-hits).

In these contests one has few options. He can maintain an aloof dignity, and methodically pound his nails while wrapped in a cloak of stately standoffishness, or he can sneak in an over-sized framing hammer, locate the softest spruce lumber, and spit on his nail before setting it. Watch your thumb!

Like all framing, floor joists and ceiling joists must have provisions for openings (basement or cold cellar stairs), and interior partition walls. Stairwells should show double headers and trimmers (see figure 8-3), and allow adequate space for the finish trim on the stairway. Should tail joists be necessary, they should be butt-nailed to the header. When there are more than four tail joists, it is a good idea to supplement the butt-nailing with metal anchors. The latter are available at most building supply stores.

Interior partitions that run contrary to the direction of the joists do not need special joist arrangements, but those paralleling the joist direction need framing attention. Figure 8-4 shows how floor joists and ceiling joists should be block-spaced so as to provide for nailing surfaces for flooring or a ceiling covering, and also to provide for the installation of plumbing or electrical work.

Blocking, in the form of solid wood pieces or diagonally-placed wood pieces, is a traditional method for assuring that the joists will not twist as they cure, but recent studies indicate that this blocking is unnecessary if the subflooring is properly nailed. We ignored the studies, and inserted solid blocking between floor joists having spans that extended more than 10 feet. Our feeling was that blocking helped to make the floor more rigid.

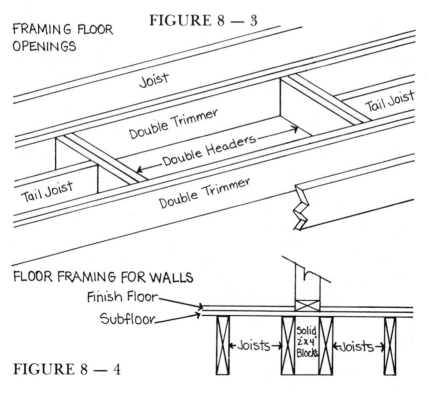

FRAMING FLOOR OPENINGS

FIGURE 8 — 3

Joist

Double Trimmer

Tail Joist

Double Headers

Tail Joist

Double Trimmer

FLOOR FRAMING FOR WALLS

Finish Floor

Subfloor

←Joists→

Solid 2x4 Blocks

←Joists→

FIGURE 8 — 4

Flooring

Rigidity is a desirable factor in most floors. If properly joisted and surfaced, a floor should never flex more than 1/360th of its span. This national code requirement amounts to about ½ inch in 15 feet. The flooring you use, and the way that it is put on contributes greatly to a floor's rigidity. Subflooring, that "skin" put on directly over the joists, can consist of one inch boards, or it can be of ⅝" or ¾" plywood. Laminated, exterior grade plywood, while stronger than comparable thicknesses of common boards, is generally more expensive. Moreover, where a warm floor or acoustical conditions are a factor, plywood is inferior to boards.

If you choose to make your subfloor of wood, you can put it on so that the boards are at right angles to the joists, or you can nail it on diagonally. The former is faster, and involves less waste, but the latter is stronger, and allows you to run narrow-boarded finish flooring in either direction. The boards to be used for subflooring may be square on their edges, shiplapped, or tongue-and-grooved. Shiplapping and tongue-and-grooving the subfloor (or the finish floor) is more desirable because their overlap keeps unwanted cracks from developing as the wood cures, but these choices are more expensive than square-edged boards.

Subflooring commonly comes in six, eight, or 10-inch widths, and when put down should be nailed with 8d or 10d common nails — two nails per six inch board at each joist, and three per 8 and 10-inch board.

There are too many options in finish flooring to allow us to discuss them all here. Synthetics, carpets, tiling and linoleums are just a few that vie with traditional wood floors for the builder's dollar. The latter also present a gamut of choices ranging from random-width, soft-wood boards to evenly-sized, narrow hardwood ones.

If the builder chooses to use a wood covering as his finish floor, it is a good idea to precede this covering with a layer of aluminum builder's foil (or tarred building paper). This layer prevents moisture damage, and inhibits insect infestation.

A final warning note should be added here about using salvaged wood flooring: Don't! Used flooring of the tongue-and-groove or shiplap variety can never have their worn or sanded surfaces re-matched to anything resembling a smooth surface.

Interior Walls

In the few years since we built our house, we have talked stone house building with many prospective builders. It seems to us that each of these builders, actual and prospective, has a different idea of how to finish the inside of the exterior walls. Proposed ideas we have discussed included: plain concrete that was molded and sculpted with butcher paper (an old Bernard Maybeck trick), concrete "sandwiched foam" walls, sprayed foam walls (either sprayed between studs, or directly applied to the inside of the concrete wall), and old style plaster that is spread over the face of the concrete.

While the plaster method was used by Edward-Flagg, Frazier Peters and other pioneers of the slipformed house, it is not a practical nor enduring finish. Other methods we have discussed have turned about the idea of using furring strips (2" — 3" strips of wood used vertically or horizontally as nailers and spacers). Builders contemplating the use of these furring strips, intended them to support a finished wall that would house solar, geothermal or other alternative heating systems.

Most of these ideas revolved around the problems of insulating a structure, and they are a regional statement reflecting our cold New England location. In the absence of knowing at first hand how all of these ideas have worked out, we hesitate to commend them to the reader, but two methods that we *can* recommend for cold weather houses are:

1) The common stud wall. This is the method that we used, and it entailed studs 16 inches on center with aluminum-faced insulation stapled between the studs. We chose to finish this wall off with wallboard (gypsum) and cherry wood. Our reasons for employing this method for our interior wall were quite pragmatic — the materials were easily available, and they work.

2) The "Rex Roberts Wall". Neighbors who built their interior walls in the manner advocated by the structural engineer Rex Roberts (see bibliography) are pleased with the results. Our visits with these neighbors (visits that included some of the coldest weather that Vermont has seen) have borne their satisfactions out. This method employs furring strips to create insulative "dead air" spaces when faced with aluminum builder's foil. Figure 8-5 shows how these furring strips are applied, and you can readily see that you can add as many insulative spaces as you feel is necessary. The method is clean, and it is inexpensive.

Roofs

Of all the house framing, roofs seem to present the most difficulties for builders. The fundamental principles of roof construction are, however, quite simple. A roof must shed water, and it should be strong enough to support itself and potential snow loads under heavy gusts of wind. We elected to top our house with a common

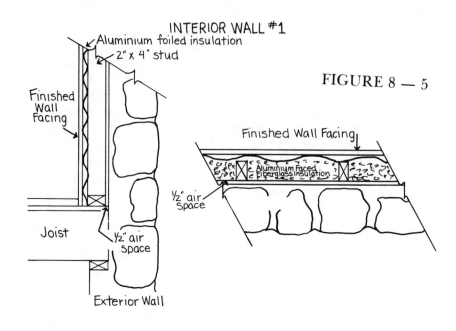

INTERIOR WALL #1

FIGURE 8 — 5

Section   Plan

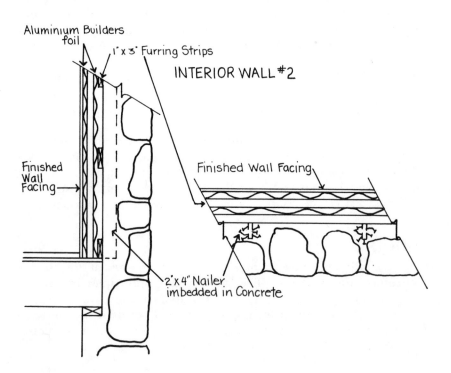

INTERIOR WALL #2

gable and valley roof, but a stone house seems to blend well with any of the four basic styles of roofs. These basic roofs are:

1) Gable — This roof has two roof slopes meeting at a ridge. The slopes may be uneven;

2) Shed — This is a single-sloped roof which is the easiest of the four to construct;

3) Hip — Having four or more sloping facets that meet at the center, the hip roof is constructed of rafters that extend diagonally from the corners to the center. Jack rafters (partial rafters) then connect into the corner rafters;

4) Gable and Valley — As the name implies, this is a combination of two gable roofs intersecting each other. The valley is created where the two slopes meet.

Each of these four basic roofs possesses four components: 1) Underpinnings — which consist of walls (exterior or partition), beams (and sub-beams), purlins, and posts; 2) Rafters — This is the framework of the roof, and it rests on the underpinnings; 3) Roof Skin — Like subflooring, the roof skin can be plywood or one inch boards. The latter are sometimes spaced to accommodate the nailing of shingles or metal roofing; 4) Roofing Material — This is the final weathering surface of the roof.

The simplest roof to construct is a shed roof that has rafters extending from one exterior wall to another. Anything beyond this is an added complication. It has been our experience, however, that few builders choose to design their houses with this appealingly simple roof construction.

Designing a house requiring a rafter length that is impossible or impractical to have a sawyer cut, requires the builder to fabricate one to suit his needs. And where the rafter is fabricated, it is necessary to include a supporting member, such as a beam or purlin. The latter may be of such length that it, in turn, requires support from a post or an interior partition wall.

From this you can see that the rafter is an important and determining factor in the construction of a roof. Table 8-1 shows the relative length to spacing relationship of the rafter when using top quality wood. These relationships are of vital importance in planning the style and components of any roof.

The footnote at the bottom of table 8-1 also points out that these length/spacing figures are for roofs having slopes (pitches) greater than 3 in 12. By 3 in 12 the carpenter means that the roof rises three inches for every foot of horizontal distance covered.

A framing square is a handy tool to have around any building site, but it really comes into its own when marking or sawing roof rafters.

TABLE 8 — 6
RAFTER SPAN, CONVERSION DIAGRAM

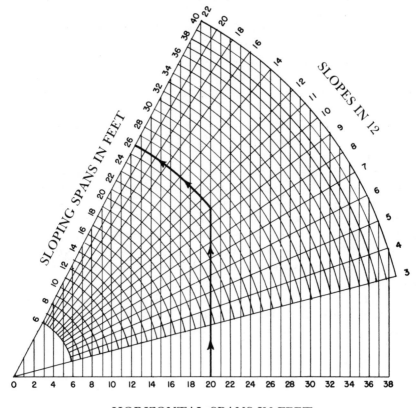

HORIZONTAL SPANS IN FEET

To find the rafter span when its horizontal span and slope are known, follow the vertical line from the horizontal span to its intersection with the radial line of the slope. From the intersection follow the curve line to the sloping span. The diagram also may be used to determine the horizontal span when the sloping span and slope are known, or to determine the slope when the sloping and horizontal spans are known.

Example: For a horizontal span of 20 feet and a slope of 10 in 12, the sloping span of the rafter is read directly from the diagram as 26 feet.

(from *FHA Minimum Property Standards* for one and two living units, 1966)

Some framing squares come complete with etched length/side cut tables, and some with length of rafter tables. Rather than relying on the reader finding his information there, we are including the information in the form of a simple diagram (table 8-6). This diagram shows how to determine rafter length from exterior wall plate to ridge. Extra lengths for eaves or overlooks should be added to the rafter length indicated on the table.

Rafters are usually erected in one of two ways: They are pre-assembled into trusses and erected as shown in figure 8-7; or they are put up singly, by nailing each rafter to a ridge board. Pre-assembled trusses should be carefully cut and assembled from a pattern. When ready to erect on the walls, they will have considerable heft (especially when there is a long span), and it pays to have a few extra hands around for the muscle work. To save on frayed tempers amongst your helpers, lay out the location of each truss on the wall plate before you begin.

Although we had good friends willing to help, we chose to put our roof up rafter by rafter (see figure 8-8). We first braced the ridge-board (a 1x10) into place, and carefully leveled it. Then we sawed pattern rafters from 2x8 dimension stock, and used these patterns to lay out and saw all of our rafters at one time (excepting jack rafters).

Jack rafters are the short ones that span the area from the valley to the ridge board, and they each require special, compound cuts. As with preassembled trusses, it is a good idea to lay out the entire raftering job on the wall plate beforehand. With this method you should also mark the ridge board. Obviously the marks on the wall plate and ridge board should match. You can be sure of this by marking the ridge board before erecting it, and laying it directly on the wall plate to check that the marks register.

The pattern that we used for our rafters included a properly-angled inset (called a birdsbeak), which we determined by transferring our roof pitch to the rafter pattern. We did this with our framing square. The purpose of the birdsbeak was to provide a stable seat for each rafter atop the wall plate. Below the birdsbeak, we allowed a generous length for our eave overhang. Once this pattern was cut, it was a simple matter to trace its outline on each rafter for the final sawing. See figure 8-9.

Putting the first rafters into place was the most difficult part of the task. We began by leaning several of the precut rafters against the outer walls so that we would not have to interrupt the framing operation with fetch-and-carry details. Then, with one of us on the wall plate and the other on a ladder, we fitted each one into place. We butt-nailed the rafter at the ridge board, and toe nailed it at the wall plate.

As we progressed with the raftering, we found that we had to erect rafters on both sides of the roof in order not to distort the straight line of

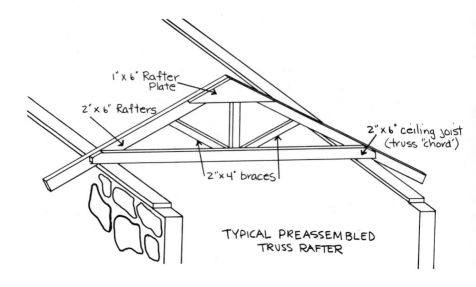

1" x 6" Rafter Plate

2" x 6" Rafters

2" x 6" ceiling joist (truss "chord")

2" x 4" braces

TYPICAL PREASSEMBLED
TRUSS RAFTER

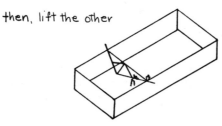

First, lift one side
to top of wall

METHOD FOR ERECTING TRUSS
RAFTERS

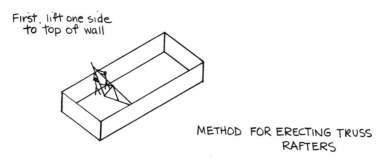

then, lift the other

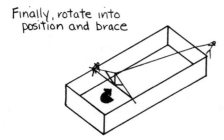

Finally, rotate into
position and brace

FIGURE 8 — 7

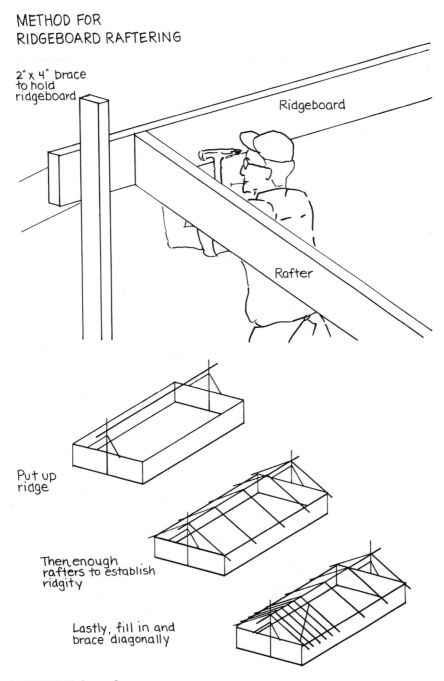

METHOD FOR
RIDGEBOARD RAFTERING

2" x 4" brace
to hold
ridgeboard

Ridgeboard

Rafter

Put up
ridge

Then, enough
rafters to establish
ridgity

Lastly, fill in and
brace diagonally

FIGURE 8 — 8

FIGURE 8 — 9

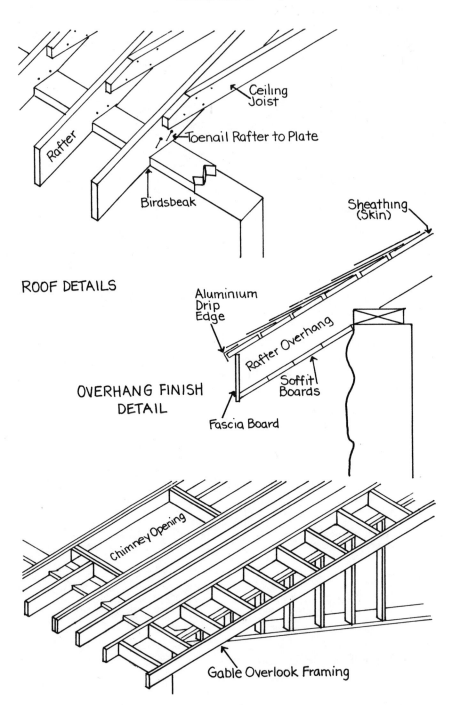

Ceiling Joist

Rafter

Toenail Rafter to Plate

Birdsbeak

Sheathing (Skin)

ROOF DETAILS

Aluminium Drip Edge

Rafter Overhang

OVERHANG FINISH DETAIL

Soffit Boards

Fascia Board

Chimney Opening

Gable Overlook Framing

the ridge board. Once we had a sprinkling of the rafters in place, and a diagonally placed board nailed across their upper surfaces, the framing took on some rigidity, and the framing progressed more smoothly.

At this stage we interrupted the raftering to nail on our ceiling joists. On the latter we strung planks from which we could more easily work to finish the raftering.

Special framing problems occurred when considering the opening we would leave for our chimney, and how we would frame the overhangs at our gable ends (overlooks). Our solution was to treat the chimney opening in the same manner as openings in the floor joisting. We included double headers and trimmers, and took care to allow for a two-inch margin between the rafters and the stone facings of the chimney. We later bridged this margin with flashing and counter flashing. See figure 8-9.

In constructing the gable overlooks, we butt-nailed short rafters to the common rafters (at right angles to them), and concluded the roof with a final rafter that omitted the birdsbeak cut. As a rule-of-thumb, it is wise not to extend overlook rafters beyond the end wall more than they reach inside the wall plate. These rafters are cantilevered, and their joint with the common rafter should be supplemented with anchor plates.

After the complications of raftering, putting on the roof skin (sheathing) was a snap. We had decided to finish our roof with asphalt shingles, and we therefore made our sheathing solid — to accommodate the nailing. We put this sheathing on in the same manner as we did the subfloor, and, as on the subfloor, we preceded the final skin with a layer of asphalt paper.

"Let me live in my house by the side of the road
And be a friend to man."

Sam Walter Foss

"This architecture we call organic is an architecture
upon which true American Society will eventually be
based if we survive at all. An architecture upon and within
which the common man is given freedom to realize his
potentialities as an individual — himself unique, creative,
free."

Frank L. Wright

The Fireplace

Chapter 9

A fireplace is an inefficient heater, a fire hazard, a source of smelly gases, a dust catcher, and a general domestic nuisance. Its chimney sucks already-heated air out of a room like a vacuum cleaner, its ash dump always seems full, and it is possessed of a perverse individuality that sees it react to every ounce of barometric pressure, knot of wind or degree of temperature. You can never cut enough wood to satisfy its voracious appetite, and you must maintain an enormous larder of twigs, shavings, paper and assorted sizes of seasoned wood to satiate its finicky palate. The fireplace loads the ground with too much weight, it tends to settle unevenly, and its building requires an inordinate amount of construction time, money and patience. A fireplace is like a child: It is difficult to conceive, expensive to feed, and has trouble supporting itself in old age. In short, there is no *rational* reason for wanting a fireplace.

But what is a stone house without one? It is tempting to launch into an anthropological justification that no stone house would be complete without a stone fireplace, but we will constrain ourselves.

It would be fun, though, to do a Kinsey-like study of how many love affairs were consummated on a rug before the open flames of a fireplace. We suspect that there was a lot more contemplation than consummation, because one of the parties was always getting up to stir the fire, add another log or just to tinker with the damper. At any rate, we decided that we wanted a stone fireplace in our stone house.

We were under the misapprehension that we could depend on it for a portion of our heat, and so we bent some of our construction efforts towards a *more* efficient style of fireplace. The operative word here is *more*.

We chose a steel "modified", heat-circulating fireplace unit which we would put into place on our hearth, and facade with stone. This unit boasts a two-walled firebox, with cold air inlets and hot air outlets. The idea is that cold air is drawn into the hollow space between the two steel walls at the base of the unit. It is then heated by the hot inner wall (adjacent to the flame), and rises into a convective current which flows out of the upper hot air outlets. The theory works, and the unit *is* more efficient than a non-circulating fireplace. But the whole thing does not hold a candle to a wood stove or a wood-burning furnace for honest heating.

Buying a ready-built steel unit like ours is one option for builders. You can get them with or without the double wall circulation, and they offer the advantages of having proven workable parts that are built in. These parts include the firebox, damper, throat, smoke shelf and smoke chamber. If installed properly they do not smoke, and they are relatively easy to install. Hence they save considerable time — an important factor for builders who have a short building season.

However, the reader may well want to consider other options before committing himself. If time allows and/or special skills (such as welding) warrant, you may want to entertain the idea of making your own steel unit. We have had occasion to prop our feet on the hearths of some fine homemade fireplaces. Most of these have been made from "boiler plate" steel, and one was cleverly fashioned from two old, differently-sized, discarded steam boilers.

With an arc welder, and considerable ingenuity, this builder wedded them together, one inside the other. He then sealed the space between them to create a convective circulating area for hot air. Through the top of the unit he welded a smoke chamber and shelf, which assured him a good draw and the elimination of undesirable downdrafts.

It is the creation of the smoke shelf and chamber that seems to confound most builders who undertake the project of building their own. There are considerable variations in how they may be constructed. With steel this is less of a problem than it is with the more conventional brick materials. Care should be exercised in choosing relational measurements to assure a fireplace's workability. You will find it difficult to correct masonry fireplace errors.

Figure 9-1 shows a conventional homemade masonry fireplace, and you can see that part of the difficulty in constructing the chamber and shelf arises from the confusion of seeing the area in question in two dimensions. The depth of the shelf (from front to back) may vary from six to 16 inches, depending upon the firebox depth. The back of the shelf should be in the same vertical plane as the back of the flue.

Using bricks for the interior of the firebox, throat and smoke chamber allows the builder to make an orderly, progressional taper of the smoke chamber. This inside wall can be overlaid on the outside with fieldstone. Fieldstones can also be used in constructing the smoke chamber, but because of their irregularity, the corbelling is more

FIGURE 9 — 1

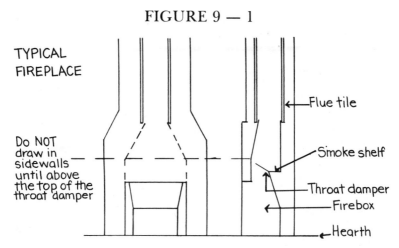

TYPICAL
FIREPLACE

Do NOT
draw in
sidewalls
until above
the top of the
throat damper

Flue tile

Smoke shelf

Throat damper

Firebox

Hearth

difficult to accomplish.

It is *not* a good idea to face the interior of the firebox and throat with stone. Stone tends to crack with sudden heating and cooling, and some that have retained sealed-in moisture have been known to explode like small fragmentation bombs when first subjected to a fire. The best material to use for these hot spots is firebrick.

We faced our fireplace with stone, placing the latter adjacent to, but not against, the modified steel form unit. We kept the stone away from the steel shell because of the flexing expansion and contraction of the metal when it is heated and subsequently cools. These units come with an adequate set of installation instructions, but they neglect some of the more practical problems you will encounter — such as how you keep the masonry away from the metal when corbelling inward over the throat and smoke chamber.

In our fireplaces, we corbelled that area with carefully-chosen stones. This was a painfully long process, and since that time we have helped with other fireplaces where the spacing was achieved by laying the masonry smack up against the unit that has been pre-wrapped in a blanket of fiberglass batts. We have followed the aging of these fireplaces, and to date have observed no cracks. For all intents and purposes, it would appear to us that the spacing provided by these batts gives a viable, time-saving and labor-saving method.

Figure 9-2 shows the construction methods we employed on our fireplaces. Back-to-back, our fireplaces shared a common hearth (cantilevered), and a common four-inch wythe (separating wall). To lessen the weight of the overall fireblock, we hollowed the walls whenever possible, and filled the resultant cavities with cement clinkers, broken cement block, and shards of broken brick. All of the stonework of

FIGURE 9 — 2

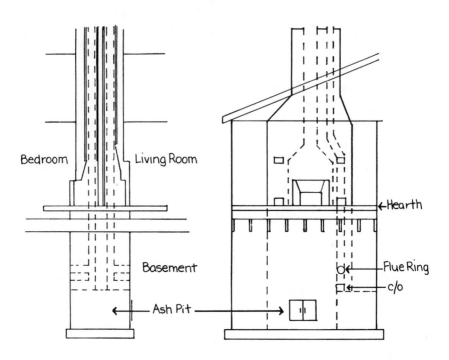

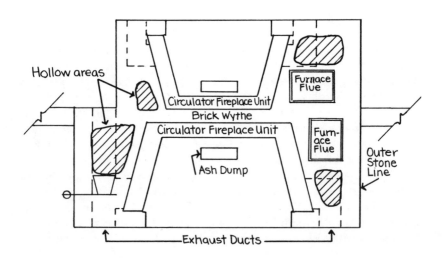

these fireplaces was hand laid (no forms were used), and stone selection was a critical factor in the process.

Planning a modified or homemade fireplace is not difficult to do once you have determined where it is to be. In placing the fireplace you should bear in mind that, unlike furniture, it is immovable once laid up. Care is necessary in the planning stage to make sure that the place you have chosen to put it does not conflict with critical framing of the floors or the roof. Trying to frame a chimney in a roof valley is like trying to tailor a sweater for an octopus.

As a practical matter you ought to decide beforehand whether you intend to burn wood or coal. If you plan on burning wood, you should make up your mind what kind of wood (hard or soft) you will use, and you might even decide what lengths of sticks are optimal for you to cut or buy.

These decisions bear on the size of opening and depth of firebox you should utilize. A 30-inch wide fireplace opening can, for example, comfortably accommodate a 24-inch stick of wood, and this is a practical division of common four foot cordwood. Coal, on the other hand, can use a narrower opening, and the depth of the firebox can be reduced.

Shallower fireboxes are more efficient for heat production, but they hold correspondingly less fuel, and, in the case of wood, they bring the fire and potential fire-starting sparks or firebrands closer to the room. We have seen fireboxes as shallow as 12 inches that worked quite well. In any case, for safety's sake, a close-meshed screen and fire dogs are necessary accessories.

The height/width ratios for fireplace openings for fireplaces that you make yourself are given in table 9-3. As a general rule, you can assume that the higher the fireplace opening, the greater is the possibility of a smoky fireplace. The old-fashioned colonial walk-in kinds of fireplaces, while picturesque, were typically smoky and grossly inefficient heaters.

But heating efficiency, as was pointed out above, is not always a criterion with fireplaces.

It is an established fact that today most fireplaces are built as part of an outer wall. This practice results in the loss of a good part of the potential radiant heat produced on the back side of the unit. We chose to build our fireplace block in the center of our house, hoping to capture as much of this radiation as possible.

While we have not calculated the effects of this planning, we do notice that the stones of the fireplace retain considerable heat long after the fire has gone out. Our wood furnace flue also runs through this fireplace block, and we reap the radiant benefits of this heat whether or not we are presently using the fireplaces.

TABLE 9 — 3

Recommended Dimensions for Fireplaces. Size of Flue Lining Required.

Size of Fireplace Opening			Minimum width of back wall	Height of vertical back wall	Height of inclined back wall	Size of flue lining required	
Width	Height	Depth				Standard rectangular (outside dimensions)	Standard round (inside diameter)
Inches	Inches	Inches	Inches	Inches	Inches	Inches	Inches
24	24	16-18	14	14	16	8½ x 13	10
28	24	16-18	14	14	16	8½ x 13	10
30	28-30	16-18	16	14	18	8½ x 13	10
36	28-30	16-18	22	14	18	8½ x 13	12
42	28-32	16-18	28	14	18	13 x 13	12
48	32	18-20	32	14	24	13 x 13	15
54	36	18-20	36	14	28	13 x 18	15
60	36	18-20	44	14	28	13 x 18	15
54	40	20-22	36	17	29	13 x 18	15
60	40	20-22	42	17	30	18 x 18	18
66	40	20-22	44	17	30	18 x 18	18
72	40	22-28	51	17	30	18 x 18	18

Abstracted from: *Fireplaces and Chimneys,* USDA Farmers' Bulletin No. 1889.

In constructing a homemade fireplace, you should be careful to follow the measurements set out in table 9-3, and when slanting the throat of the fireplace, you should never allow the volume of the throat to be less than a comparable length of the size of flue tile indicated for your size of fireplace opening. The throat should start to slant inward from the firebox sides no less than six inches above the bottom of the lintel, and should be interrupted (see figure 9-1) with a damper. The latter is invaluable in cold climates for controlling draft, keeping out insects, and sealing the chimney off from heat leaks when the fireplace is not in use. Dampers are available from most building supply stores, and sometimes they come complete with throat assembly.

All fireplaces should be based on a generous hearth — (where else would the dog or cat sleep?). The purpose of a hearth is that of providing a base for the fireplace to sit on, and in its extension it provides a safety apron which prevents sparks or firebrands from spilling out on the floor. This extension should be at least two feet wider than the fireplace opening, and should extend 20 to 24 inches beyond the face of the fireplace opening. Whether the hearth is raised or not is more a matter of personal preference than a question of heating efficiency. We chose to elevate and cantilever our hearth, and we surfaced it with flat stones.

All hearths should be made of a non-flammable material, and it is a good idea to make that part of the hearth on which the fires will be built of firebrick that is mortared with fireclay.

Unless you plan on forming a reinforced arch over the opening to your fireplace, you should use a steel lintel to support the masonry. We employed a length of angle iron (4" x 4" x ⅜"), which we cut of sufficient length to allow us to project it six inches into the adjoining jambs. This size of angle iron should be adequate for openings up to four feet in width. Hand-laying the stone on this lintel is a ticklish business, and calls for very careful stone selection to assure that the steel lintel is not visible, and that each stone is secure in its seat.

It is impossible to convey all of the subtle techniques employed in hand-laying stone in a fireplace wall — which probably accounts for the fact that there are so few books written on the subject. Conventional stone masonry is largely a matter of experience and "feel". We found, for example, that the finished look of our fireplaces improved as we progressed, and this can be construed as advice for the builder to work out his own techniques on some other, less critical structure — say on a retaining wall.

There are some fundamental principles for, and ways of going about, hand-laying fieldstone that will help. Many of these suggestions have already been discussed with the slipforming process, but we are repeating them here for continuity.

● Before you begin, you should decide how you want the finished product to appear. Do you want the fireplace to exude a sense of massiveness? delicacy? patterns? archwork over the opening? The answers to these questions will dictate your accumulation of selectable stones. They will also indicate the method of pointing you will employ (*i.e.*, recessed pointing, or flush pointing — see figure 9-4).

If you decide that you want to feature one or more special stones (strikingly beautiful or unusual ones), you will want to have some mental picture of how you will achieve the effect you are after.

Over our living room fireplace we had a particular stone that we wanted to feature. The stone was not especially beautiful, nor was it unusual, except for two long striations that marred its otherwise plain schist face. These silent scars recorded the story of our first day's effort at tilling the stony soils of our Vermont farm. They are the spoor of our plow, and in their etched tracks they record the story of a man who was in too much of a hurry. It was the hurry, not the stone, that broke the plow point.

We featured this large story-stone by the simple design of placing it in a vertical position, and then surrounding it with smaller stones that were in horizontal planes. Now in solitary splendor above our mantel,

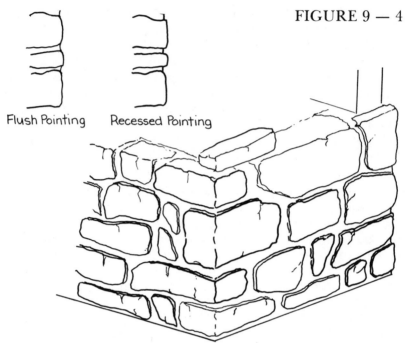

FIGURE 9 — 4

Flush Pointing   Recessed Pointing

the stone does not fire the imagination of visitors, but it has a special significance for us.

 ● In hand-laying a fireplace it seems that there is a need for an inordinate amount of quoins (cornerstones). We made it a practice to search for them whenever we came upon a new stone pile, and when we gathered them we set most of them aside in the fireplace pile. Figure 9-4 shows how to lay up a well-knit corner, and you can see from this illustration that the basic rule of ONE OVER TWO, AND TWO OVER ONE applies to hand laying as well as slipform work.

 ● To bind the stones together we used a mortar consisting of three parts screened sand to one part of portland cement. We used no lime, finding that this mixture gave us a mortar with enough body that it would stand by itself when placed, and yet move under gentle tapping. When mixed properly, this mortar has a "buttery" consistency that makes a nice finish when stroked with a smoothing trowel.

 ● Before applying the mortar under each stone, we first made a dry fitting to ascertain just how the stone would sit. This routine also gave us a chance to step back and assess how this particular stone contributed to the overall picture or design that we were striving for. If the stone fit and seemed stable, we removed it and prepared its bed. The preparation included a one to two inch bed of mortar. Then, buttering the conjunctive side of the stone (the side of the stone that would butt against its neighbor), we gently tapped it into position. The tapping seems, to the onlooker, almost ritualistic, but its functional purpose is to eliminate any air pockets that might have formed in the mortar.

All of this seems simple enough, but in practice you will find that the stone either won't fit the space, or that it teeters so precariously in the dry laying that you know that it will not remain in place until the mortar cures. Finding the right sized stone for a specific place is largely a matter of having a large enough selection. As we worked on our fireplace, we found ourselves constantly looking two or even three stones ahead in the laying. Stones that teeter can be adapted in one or two ways: You can cut or chip them to the desired configuration, or you can wedge them into a stable position. For wedging purposes, we kept stone chips and shards of all sizes handy. After a stone chip was placed in a position to steady the problem stone, we laid the bed of mortar right over it.

● Making a fireplace with hollow cavities usually means that the stone wall is no thicker than the stones chosen for the face. In our case, this meant a wall of about 12 inches, and because the in-process, uncured strength of this wall depended exclusively upon the gravitational force of the overlapping stones acting upon each other, we never erected more than four feet of vertical rise in the wall per day. Filling the hollow cavity with clinkers, shards and cinders as you go *does* help the stability, but trying to stretch a day's progress to "just one more course" can result in a Jericho-style disaster.

FIREPLACE FOOTINGS

Ash pits and cleanouts are not absolutely essential items for fireplaces, but good footings are. When completed, a modest fireplace and chimney can weigh in the neighborhood of 20 tons. This kind of weight makes it essential that the footing be made large enough to support the fireplace masonry. Too often careless builders treat the fireplace footings as though they were common wall footings. This carelessness results in differential settling that sees the fireplace block settle faster than the rest of the house. In one case, where a builder mistakenly moored the fireplace block to structural members of his house, we saw the warped floors and swaybacked roof that resulted when his overweighted footing succumbed to its load.

Ground loading for fireplace footings should be calculated in the same way as outer walls (see Chapter 4). If an ash pit, septum walls or hearth extensions are to be built into the block, their weights must be made part of your calculations. We figured the weight of our fireplace block to. be somewhere in the neighborhood of 60 tons. To support that weight, we poured a 12-inch thick, reinforced pad under the entire block, and extended it six inches on all sides of the perimeter. This gave us a ground loading per square foot that was approximately the same as that for our walls (around 1800 pounds per square foot).

Figure 9-5 shows how we constructed our footings, ash pit and hearth. In order to accommodate the other flues (wood and oil furnaces), we inserted flue rings and cleanout openings between our slipforms at the appropriate spots in the ash pit walls. The flue rings were put in directly, and the cleanout openings were framed in wood in the same manner as for wall ventilators (see Chapter 7).

Cleanouts should be located directly below their appropriate flues, and the top of each cleanout door should be at least 12 inches below the bottom of the flue ring. This is to allow for easier cleaning of collected soot, and to provide for the chimney flue tile extending at least eight inches below the bottom of the flue ring.

Forming and pouring the septum walls shown in figure 9-5 was difficult, but we wanted to confine the ash pit area for ease of cleaning, and to provide extra support for the fireplace walls that began atop the hearth. Our ash pit doors were cast iron sugar arch doors, which we salvaged from a neighbor's abandoned sugar house. To support our cantilevered hearth, we placed extra vertical reinforcing bars in the ash pit walls, and bent them over for transition into the hearth.

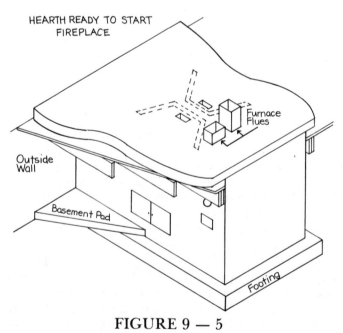

FIGURE 9 — 5

In the pouring of the hearth we included all manner of extra salvaged steel reinforcing. We put in three lengths of railroad track, 36 feet of four-foot-high sheep fencing, two tractor axles and quantities of electric fence wire. From the beginning, the pouring of the hearth involved several steps. We braced and scaffolded forms into the gaping openings over the ash pit, and formed openings for the ash dump doors in the hearth. We then placed the chimney tiles up from the furnace flue rings, and formed the cantilevered section of our hearth. Finally, we poured the hearth itself (in one operation), and finished it off with a flagstoned surface.

After curing, we dismantled and removed the staging and forms from the ash pit (taking them out through the ash pit door was quite a feat), and knocked out the forms and staging from beneath the cantilevered section of hearth. This gave us one continuous, stone-faced hearth with two apertures into which we cemented ash dump doors, and two chimney flues that would exit smoke from our furnaces.

THE CHIMNEY

Chimney flue tiles are best included in fireplace plans. They are relatively simple to clean and maintain, and they allow a more efficient circulation of air and gases. Most flue tiles are made of vitrified fire clay, and they come in rectangular or round shapes. The latter are more efficient, but they are not always stocked in building supply stores. It is a good idea when estimating and buying flue tile to acquire your order all at one time. Batch and foundry dimensional differences occur, and while usually small, they are irritating to contend with.

While at the building supply store, you may want to acquire a carbide blade for your circular or sabre saws. These tools make cutting a flue tile for offsetting or for the insertion of smoke pipe a much easier task. The alternative to this is to fill the tile with wet sand, and then to cut it with a hammer and a chisel.

Before beginning to erect flue tile, it is wise to rig a bag filled with sand or sawdust to hang down in the tile. Mortar and tools that are inevitably dropped down into the flue while you are working will be caught on the bag, and may then be extracted at your option. But do not make the same mistake we did of using rotten rope to secure the bag. It took us two hours of probing, fishing and swearing to extract our wedged sand bag.

Flue joints, as used here, are the unions of chimney flue tiles — not hospital dispensaries. Where two or more flues occupy the same chimney, the flue joints should be staggered so that the joints do not occur in the same horizontal plane. This calls for some juggling, as each section of flue lining is set in place ahead of its section of chimney wall or wythe. The process of joining flue tiles is simple enough. You lay a bed of mortar on the rim of the preceding tile, and set the new tile onto it. The simple nature of the job ends there. Flue tiles are heavy, cumbersome, breakable and expensive. You must work with them at heights, and the tiles used at the top of the chimney weigh just as much as the ones at the fireplace level. They just seem heavier.

Once put into place, the tile should be plumbed with a level — unless you are deliberately offsetting a flue. (Offsets, by the way, should never angle at more than 45 degrees from vertical.) Tapping on a wood block at the top of the flue will usually bring the flue into plumb, but it often results in chunks of fresh mortar being dislodged from the union. More often than not this leaves unwanted holes, and they must be filled. After you are sure that the joint is whole (we checked for holes by dropping a cord light into the flue and inspecting the outside of the seam for light leaks), you must smooth the inside of the seam. This smoothing will prevent shelfs from forming as mortar drops from subsequent seams, and will make a surface that will inhibit potentially dangerous accumulations of creosote.

In most masonry chimneys (those with walls less than eight inches thick) there should be a space between the tiles and the chimney walls. This is easier said than done, when hand laying fieldstone.

We found that the technique of wrapping the tiles with fiberglass batts (as was suggested with modified steel fireplace units), simplified the procedure. We then laid our stone up against the fiberglass-buffered chimney tiles. The fiberglass also provides needed insulation for the flues as they emerge above the roofline.

As was pointed out above, chimneys may contain more than one flue, and to provide for how these flues are arranged you should have one drawing showing your chimney layout at the roofline. We planned a separate flue for each heating unit (this is a building code stipulation in many areas), and we separated the flues from each other with a four-inch brick wythe.

It was a satisfying experience for us to vindicate our planning, as we brought our chimney up between the roof rafters. As was planned, the stonework and flues fit properly, and we found that we easily met the requisite two-inch gap between masonry and wood.

The moment that the masonry clears the roof, you must begin preparations for counter flashing the juncture of masonry and roof. The normal procedure is to build flashing into the roof surface. This roof (or base) flashing is then bent upward alongside the chimney, and the counter flashing, which is bonded into the mortar joints of the chimney, is then lapped down (countered over) the base flashing.

CHIMNEY PLAN AS IT EXITS ROOF

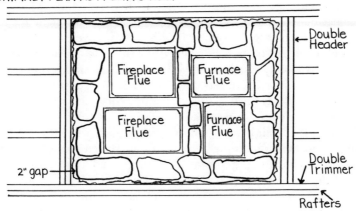

FIGURE 9 — 6

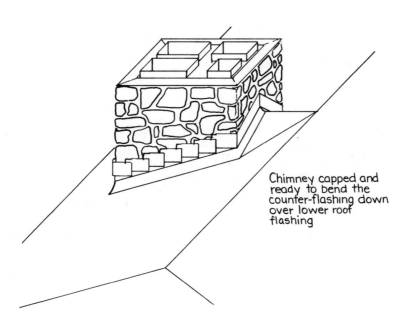

Chimney capped and
ready to bend the
counter-flashing down
over lower roof
flashing

However, unlike brick masonry, fieldstone does not normally provide evenly coursed joints into which you can place your stepladdering counter flashing. We resolved this problem by using bricks below each sheet of counter flashing. This gave us a reliably even base, and the bricks do not show when the flashing is bent into place.

Flashing may also have to be provided for a cricket on the upper side of the chimney. Crickets are dormer-like extensions above the chimney that are designed to shed unwanted snow and ice loads in cold climes. See figure 9-6 for details.

The final height of a chimney depends on several factors, all having to do with obtaining a clear, unobstructed draft. Under most circumstances, raising the chimney to a point two feet above the highest point of the adjacent ridge is sufficient. To finish off the chimney, you should allow the flue lining to project at least four inches above the final cement cap (see figure 9-6). The latter should be at least two inches thick, and it should drain water away from the flue tiles.

You may wish to hood your chimney to keep out weather and/or unwanted wind eddies. For the most part, hooding is not recommended because it tends to interfere with a proper draft. When building, we were unsure on this point, and so we made provision in the final stone laying and capping for the possibility of later adding a hood. But as of this writing we have not felt it necessary.

"The most tangible of all mysteries — fire."

Leigh Hunt

* * * *

"How can I turn from any fire
On any man's hearthstone?
I know the wonder and desire
That went to build my own."

Rudyard Kipling
The Fires

Epilogue

*N*ow as we finish this book, we wonder what purpose the writing will serve. Will it influence the reader to build of stone? If, as we suspect, the reader was already predisposed towards the medium, will the writing give him the nuts-and-bolts information that he needs to build his own slipformed stone house? We would like to think that it will, but, in truth, we would hope for more.

For example, we would like to leave the reader with a bit of country philosophy — that of "makin' do." When used hereabouts, the expression implies a lot more than it says. Folks who are makin' do are those with limited external resources (mostly money), but who are rich in natural resources (stone, wood and common sense).

Despite these limitations (perhaps because of them) the makin' doers cheerfully cope. When we first got to know makin' do people, we noticed right off that they were different. They were independent, ingenious, independent, practical, independent, hard-working, independent, capable and independent.

We grew to like them, and then found that the fact that we were building our own stone house gave us some common ground with them. Eventually we found ourselves makin' do, and we are now proud to count ourselves among that company.

Building a stone house gives you a lot of time for reflection, and in the course of our building we gave considerable thought to the changes we saw in ourselves. From the outset we discovered

149

that our innate capacities to cope or make do had gotten rusty with disuse. Like so many others of this country, we had been hornswoggled into a passive acceptance of the "holiness" of expertise, and one of our first tasks was to rid ourselves of this wrongheadedness. Learning to do for ourselves, while sometimes frustrating, has been worth every bit of effort that it took.

At the risk of sounding presumptuous (and clichéic), we concluded that our experience was a reflection of what this country needs today. Not that everyone can or should go out and build a stone house, but everyone would do well to re-discover his own capacities to make do. This nation, founded and once-thriving in the hands of make-doers, now threatens to founder because its people have forgotten a capacity as native to them as their navels — that of coping for themselves on a first-hand basis.

As a nation our growing inclination has been to let "experts" manage our thinking (political and otherwise), grow our food and build our shelters for us. This abdication of responsibility has gotten us political scandals, profit gouging, chemically filled "wonder" foods and ticky-tacky housing.

We arrived in these straits by dint of slothful credulity. We were lazy, and we were naive. What is needed now is not non-participant skepticism, but rather a cautious and careful re-examination of basics to find where we can do without expertise. When confronted with the rote response, "that's a job for the experts," or "You should have a professional do that job," we must suspend (or at the very least, withhold) belief.

If through "expertise" and "specialization," our world has grown too complex for us to live on it, we must go about reducing that complexity. The stone house builder, by the nature of his material and the fact that he is doing it himself, is striving to simplify. He has not rejected the future — far from it. For the stone's durability in itself is a hopeful statement about the future. Nor has he buried his head ostrichlike in the sands of the past. He has chosen a natural material which he fashions into a house with his own hands. This act is life-affirming and is the essence of simplicity.

Bibliography

BOOKS

A Treatise on Masonry Construction, Ira O. Baker; John Wiley and Sons, N.Y.; 1899; 556 p.

Architectural Graphic Standards (5th ed.), C.G. Ramsey and H.R. Sleeper; John Wiley and Sons, Inc., N.Y.; 1956; 758 p.

Body Time (Physiological Rhythms and Social Stress), Gay Gaer Luce; Pantheon, N.Y.; 1971.

Design of Concrete Structures, George Winter; McGraw Hill Book Co., N.Y.; 1964; 660 p.

Engineering Properties of Rocks, I.W. Farmer; E. & F.N. Spon Ltd, London; 1968; 180 p.

Eureka!, An Illustrated History of Inventions From the Wheel to the Computer, Ed. by Edward DeBono; Thames & Hudson Ltd., London; 1974.

Frames and Arches, Valerian Leontovich; McGraw Hill Book Co.; 1959; 472 p.

Geologic Time, D.L. Eicher; Prentice Hall, N.Y.; 1968; 149 p.

Handbook of Homemade Power, (Alternative Energy Sources that You Can Use Now), Ed. by John Shuttleworth; Bantam Books, N.Y.; 1974; 374 p.

Houses of Stone, Frazier F. Peters; G.P. Putnam's Sons, N.Y.; 1933.

Living the Good Life (How to Live Sanely and Simply in a Troubled World), Helen and Scott Nearing; Schocken Books, N.Y.; 1970; 213 p.

Manual of Standard Practice for Detailing Reinforced Concrete Structures, American Concrete Institute (315-65, 4th ed.); The Institute, Detroit, Mi.; 1965; 165 p.

Modern Carpentry, Willis H. Wagner; The Goodheart-Willcox Co., Inc., S. Holland, Ill.; 1969; 480 p.

On Architecture (Selected Writings), Frank L. Wright (Ed. by Frederick Gutheim); Grosset & Dunlap Inc., N.Y.; 1959; 275 p.

Pour Yourself a House, Frazier N. Peters; McGraw Hill Book Co., N.Y.; 1949.

Rock Color Chart, E.N. Goddard, *et al;* Geologic Society of America, National Research Council (republished by G.S.A.); 1951.

Simplified Carpentry Estimating, J.D. Wilson & C.M. Rogers; Simmons-Boardman Pub. Corp., N.Y.; 1962; 320 p.

Simplified Masonry Planning and Building, J. Ralph Dalzell; McGraw-Hill Book Co., Inc., N.Y.; 1955; 362 p.

Sticks and Stones (A study of American Architecture and Civilization), Lewis Mumford; Dover Pubs. Inc., N.Y.; 1955; 238 p.

Stone Catalog, Building Stone Institute, N.Y.

Stone: Properties, Durability in Man's Environment, E.M. Winkler; Springer-Verlag, N.Y./Wein; 1973; 230 p.

The Forgotten Art of Building a Good Fireplace, Vrest Orton; Yankee Inc., Dublin, N.H.; 1969.

The Forgotten Art of Building a Stone Wall, Curtis P. Fields; Yankee Inc., Dublin, N.H.; 1971; 61 p.

The Natural House, Frank L. Wright, Horizon Press, N.Y.; 1963; 223 p.

Village Technology Handbook, VITA (Volunteers for International Technical Assistance); Schenectady, N.Y.; 1970.

Your Engineered House, Rex Roberts; M. Evans & Co., Inc., N.Y.; 1964; 237 p.

Windmills and Watermills, John Reynolds, Praeger Publishers, N.Y.; 1970.

PAMPHLETS

Electric Power From the Wind, Henry Clews; Solar Wind Pub.; 1973; 29 p.

GOVERNMENT PUBLICATIONS

BOOKS

(U.S. Government Printing Office, Superintendent of Documents, Washington, D.C., 20402.)

Carpentry and Building Construction, U.S. Dept. of the Army Technical Manual 5-460; 1960; 198 p.

Concrete and Masonry, U.S. Dept. of the Army Technical Manual 5-615; 1963; 166 p.

FHA Minimum Property Standards (For One and Two Living Units), U.S. Dept. of Housing and Urban Development /FHA; 1965 (rev.); 315 p.

Geothermal Energy, Federal Energy Administration (Task Force Report); 1974; 128 p.

Solar Cooling for Buildings (Workshop Proceedings), National Science Foundation; 1974; 231 p.

Solar Energy, Federal Energy Administration (Task Force Report); 1974; 596 p.

Wood: Colors and Kinds, U.S. Dept. of Agriculture Handbook No. 101; 1970.

Wood Handbook, U.S. Dept. of Agriculture/Forest Service, Forest Products Lab; 1964 (rev.); 528 p.

PAMPHLETS

A Primer on Water, Luna Leopold and Walter Langbein; U.S. Dept. of Interior; 1960; 50 p.

American Standard Building Code Requirements for Masonry, U.S. Dept. of Commerce/National Bureau of Standards Misc. Pub. 211; 1954; 39 p.

Farm Water Power, George Warren, U.S. Dept. of Agriculture; 1931.

Fireplaces and Chimneys, U.S. Dept. of Agriculture, Farmers' Bulletin No. 1889; 1963 (rev.); 23 p.

Fire Resistant Construction, U.S. Dept. of Agriculture, Farmers' Bulletin No. 2227; 1967; 18 p.

House Construction (How to Reduce Costs), U.S. Dept. of Agriculture, Home and Garden Bulletin No. 168; 1970 (rev.); 16 p.

Know the Soil You Build On, U.S. Dept. of Agriculture/Soil Conservation Service, Agriculture Information Bulletin 320; 1967; 13 p.

Know Your Soil, U.S. Dept. of Agriculture, Soil Conservation Service, Agriculture Information Bulletin 267; 1970 (rev.); 16 p.

Manual of Septic Tank Practice, U.S. Dept. of Health, Education & Welfare, Public Health Service, Publication 526; 1967; 30 p.

Plumbing for the Home and Farmstead, U.S. Dept. of Agriculture, Farmers' Bulletin No. 2213; 1971 (rev.); 20 p.

Preservative Treatment of Fence Posts and Farm Timbers, U.S. Dept. of Agriculture, Farmers' Bulletin No. 2049; 1967 (rev.); 33 p.

Protecting Log Cabins, Rustic Work, and Unseasoned Wood from Injurious Insects in the Eastern United States, U.S. Dept. of Agriculture, Farmers' Bulletin No. 2104; 1970 (rev.); 18 p.

Roofing Farm Buildings, U.S. Dept. of Agriculture, Farmers' Bulletin No. 2170; 1969; 28 p.

Soils and Septic Tanks, U.S. Dept. of Agriculture, Soil Conservation Service, Agriculture Information Bulletin 349; 1971; 12 p.

Use of Concrete on the Farm, U.S. Dept. of Agriculture, Farmers' Bulletin No. 2203; 1970 (rev.); 30 p.

SUGGESTED ADDITIONAL READING

Building Stone Walls, by John Vivian. 80 pp., quality paperback, $2.95. Many photographs and drawings, with everything you need to know to build beautiful and lasting stone walls.

Low-Cost Pole Building Construction, by Douglas Merrilees and Evelyn Loveday, 104 pp., quality paperback, $4.95; hardcover, $10. How to construct sturdy, inexpensive buildings.

Your Energy-Efficient House, by Anthony Adams. 128 pp., quality paperback, $4.95, hardcover, $8.95. Heavily illustrated, a perfect idea book for the homeowner who wants to cut those high heating costs.

These books are available at your bookstore, or they may be ordered directly from Garden Way Publishing, Dept. BSH, Charlotte, VT 05445. If order is less than $10, please add 60¢ postage and handling.

Index

155